Newton

and the Culture
of Newtonianism

THE CONTROL OF NATURE
Series Editors: Margaret C. Jacob and
Spencer R. Weart

PUBLISHED
SCIENTISTS AND THE DEVELOPMENT OF NUCLEAR WEAPONS
From Fission to the Limited Test Ban Treaty, 1939-1963
Lawrence Badash

EINSTEIN AND OUR WORLD
David Cassidy

NEWTON AND THE CULTURE OF NEWTONIANISM
Betty Jo Teeter Dobbs and Margaret C. Jacob

FORTHCOMING
CONTROLLING HUMAN HEREDITY
1865 to the Present
Diane B. Paul

CONTROL OF NATURE

Newton
and the Culture
of Newtonianism

Betty Jo Teeter Dobbs
and Margaret C. Jacob

HUMANITIES PRESS
NEW JERSEY

First published in 1995 by Humanities Press International, Inc.
165 First Avenue, Atlantic Highlands, New Jersey 07716

© 1995 by Betty Jo Teeter Dobbs and Margaret C. Jacob

Library of Congress Cataloging-in-Information Data
Dobbs, Betty Jo Teeter, 1930–94
    Newton and the culture of Newtonianism / Betty Jo Teeter Dobbs and
  Margaret C. Jacob.
      p.   cm.   — (The Control of Nature)
    Includes bibliographical references and index.
    ISBN 0–391–03878–8 (cloth). — ISBN 0–391–03877–X (pbk.)
    1. Newton, Isaac, Sir, 1642–1727. 2. Science—England—
  History—17th century. 3. Physics—England—History—18th century.
I. Jacob, Margaret C., 1943– II. Title. III. Series.
QC16.N7D66 1995
530'.0941'09032—dc20   94–18946
                    CIP

A catalog record for this book is available from the British Library

Humanities Press and the co-editors of this series wish to dedicate this book to the memory of Betty Jo Teeter Dobbs. Shortly after its completion, she died suddenly and unexpectedly on March 29, 1994. She is missed by all who knew her.

Contents

List of Illustrations

Series Editor's Preface

T HIS NEW SERIES of historical studies aims to enrich the understanding of the role that science and technology have played in the history of Western civilization and culture, and through that in the emerging modern world civilization. Each author has written with students and general readers, not specialists, in mind, and the volumes have been written by scholars distinguished in their particular fields. In this book one of the authors has published extensively on Newton and seventeenth-century science and alchemy, the other on eighteenth-century cultural and intellectual life and its relationship to science.

The aim of this book on Newton and the culture of Newtonianism is not to just lay out some basic historical information, which could only be a sample of the many complex developments that scholars are currently exploring. Still more this volume intends to show the chief questions and debates that engage current historical scholarship.

The current debates as presented here emphasize the "Control of Nature." While not excluding a discussion of how knowledge itself develops, how it is constructed through the interplay of research into nature with the values and beliefs of the researcher, this volume—like all the others in the series— looks primarily at how science and technology interact with economic, social, linguistic, and intellectual life, in ways that transform the relationship between human beings and nature. In every volume we are asking the student to think about how the modern world came to be invented, a world where the call for progress and the need to respect humanity and nature produce a tension, on the one hand liberating, on the other threatening to overwhelm human resources and ingenuity. The scientists whom you will meet here could not in every case have foreseen the kind of power that modern science and technology now offer. But they were also dreamers and doers—as well as shrewd promoters—who changed forever the way people view the natural world.

Finally, a word about citation style: whole books and essays are cited within brackets, e.g., [1], referring to items in each bibliography. When a page or volume and page are cited, the notation includes a colon, as [1:32] or [1:vol. 2, 13]; when many items in the bibliography are cited, each is separated by a semicolon, as [1; 2; 3]. Throughout the authors have allowed the year to begin on January 1 as it did on the Continent at that time, but not in England. Hence Newton would have written January 2, 1664; the authors have converted it into 1665.

SPENCER WEART

ix

An Eighteenth-Century Engraved Portrait of Sir Isaac Newton (1642–1727). In the possession of B. J. T. Dobbs.

Introduction

THE PEOPLE OF seventeenth-century England organized their lives by the rising and setting of the sun and by the rhythms of the seasons. Many were deeply religious; only a bare majority could sign their names to a contract, and still fewer could read and understand it. Yet out of that society at that time came two extraordinary transformations that would change the course of English, then American, and finally Western history. Parliamentary, representative forms of government developed out of a revolutionary process that began in 1640 and only ended in 1689; just as important, modern science coalesced into an organized body of knowledge, complete with an experimental method and an institutional base within the English universities and the private scientific societies. In the second transformation Isaac Newton towers above even his scientific peers—Robert Boyle, Robert Hooke, G. W. Leibniz—because his science provided an overarching framework within which all other sciences of the age in turn developed.

This book introduces the student first to Newton and then to what happened to his science as it was interpreted by his major followers. They found meanings in it that were both religious and practical. The Newtonian universe, organized by Newton's physics and celestial mechanics, permitted an entirely new approach to nature. Nature's wonders became knowable, no longer simply awesome or terrifying. Newton's *Principia* of 1687 presented the law of universal gravitation and laid the foundation for modern physics, mechanics, and astronomy. Its method, both mathematical and experimental, became the exemplar of how science should be done. Newton's *Opticks*, published in many editions (1704 and after), revealed the nature of light and, in a series of speculations offered at the end of the book, raised fundamental questions about the nature of matter. In the century of scientific work initiated by these publications, electrical experimentation began in earnest, while mathematical physics, mechanics, and mechanical engineering arose out of the framework provided by the *Principia*. Newtonian science in the form of applied mechanics in turn fed directly into the Industrial Revolution, which began in Britain during the last quarter of the eighteenth century.

1

The revolutions of the seventeenth century, out of which modern industrial and democratic society evolved, were both conceptual and political. While contributing decisively to the first, Newton lived through the second, and historians have long speculated on the relationship between the birth of modern science and the seventeenth-century revolution in government. In Part 2 we too will discuss in detail what some of those connections between the two revolutions may have been. One obvious linkage lies in Protestant religious belief. The men who led the revolution against the English king, and in the process created independent parliamentary institutions, found justification for their actions in the religious conviction that God endorsed their cause. Throughout his life Newton, just like the revolutionary Puritans, sought that same God, but in his laboratory rather than on the battlefield or in small sectarian gatherings.

Nature revealed the work of the Great Mathematician, and Newton sought to know his God through it. Out of these lonely speculations arose scientific discoveries that helped to usher in an age where Newton's God seemed increasingly irrelevant. Such a world he had never intended; indeed, had he lived in the late eighteenth century, Newton would have been horrified by men who thought more about capital and machines than they thought about their Creator and His eternal decrees.

Yet the *Principia* put into print more replicable and verifiable laws of nature than had ever appeared in one book in the history of publishing. Newton's achievements have earned him a place among the half dozen or so most important thinkers in Western history. The inner workings of his mind inevitably deserve careful examination. His genius fascinates, even compels, students and scholars alike. What do we make of a man who probably worked and studied during every hour of light and well into every candlelit night? In the same day Newton could be reading theology, writing biblical criticism, probing the sacred book for hints about the end of the world, preparing a chemical experiment, reading a clandestine manuscript on alchemy, working out the mathematics of universal gravitation—to rest finally only out of exhaustion. He was suspicious, secretive, seldom humorous, indifferent to poetry and music, not friendless but never an easy companion. Yet without him it is hard to imagine the intellectual history of the century that came after him. Given Newton's importance and the complexity of his thought, he demands our exclusive attention in Part 1. Then in Part 2 we will examine the social and religious setting that gave rise to the Newtonians and to the many applications found for Newtonian science.

Isaac Newton (1642-1727)

Newton's Youth

ISAAC NEWTON WAS born on Christmas Day 1642, the premature, post-humous, and only child of an illiterate yeoman farmer of Lincoln-shire in England. Not really expected at first to live—he was later to remark that at his birth he was so small that he might have been put into a quart mug—he survived war, revolution, plague, and the seven-teenth-century pharmacopoeia to the age of 84, to be buried in West-minster Abbey (the traditional place of interment for the queens and kings of England), idolized by many of his countrymen and admired by much of the Western world.

His genius appeared more mechanical than intellectual at first: as a boy he constructed water clocks, windmills, kites, and sundials and cleverly used the force of the wind to enable himself to outjump the other boys. But, nurtured by neighboring village schools and the King's School at Grantham, his intellectual prowess and his enormous powers of concen-tration slowly became apparent. Recalled from school by his mother to learn the art of farming, he spent his time under the hedges with his books and his calculations, to the utter neglect of the life of his ancestors. Eventually a maternal uncle, a Cambridge man himself, intervened to have him returned to the school at Grantham to be prepared for Cambridge University, and Isaac went up to that venerable seat of learning in 1661, entering Trinity College. He was aged 18, a little older than most enter-ing students and probably less well prepared than many, but evidently with all his faculties ready to flower. [9; 13; 17; 66; 101]

Early University Studies

Since the great period of the translation of Greek and Arabic scien-tific literature into Latin in the eleventh, twelfth, and thirteenth centuries, core curricula in the universities of western Europe had been based on the work of Aristotle (384–322 B.C.). And even though one may see in retrospect that a number of challenges to it had developed both within and without the university world in the sixteenth and seventeenth centuries, when Newton entered Cambridge in 1661 the core curriculum was still based on the Latin or Greek texts of Aristotle and on medieval and Renaissance commentators on and expositors of Aristotelian doctrines. For about two years the young Newton applied himself to learning Aristotelian logic, ethics, rhetoric, metaphysics, and natural philosophy. [55; 98:23–44]

In natural philosophy, that would have meant that he learned that the cosmos (the entire created world) was organized around the earth at the center and was made up of nesting spheres. The first set of spheres comprised the four Aristotelian elements: earth at the center, then the spheres of water, air, and fire, filling all the volume up to the sphere of the Moon; there were no empty spaces in the Aristotelian cosmos.

The sphere of the moon was a great dividing line, with change limited to the world below the moon where motions of various sorts brought about changes of several kinds. Aristotle designated four types of change: (1) generation/corruption, as in the birth and death of a living creature; (2) qualitative, like changes in such qualities as heat, cold, dryness, and wetness; (3) quantitative, as in the enlargement of something that has the ability to expand (Aristotle's example would perhaps have been inflating a sheep's bladder; a more modern example might be blowing up a balloon); (4) change of place or "local motion"—from which our word *locomotion* is derived—as a car traveling down a highway, or in Aristotle's case more likely a man walking from his home to Aristotle's school in the center of Athens.

Above the sphere of the moon, on the other hand, were the eternal and changeless spheres of the other planets, those that (including the moon) comprised the anciently known planets for which the days of the week are named: Moon, Venus, Sun, Mars, Jupiter, and Saturn. Finally, at the edge of the closed system was the sphere of the "fixed stars" that appeared not to change at all in relation to each other, only to whirl around the earth once every 24 hours. The planets (the word means "wanderers") do appear to move in relation to each other, as well as in relation to the earth and to the fixed background of stars. But the only motion allowed to those celestial objects was a circular one, a movement always returning to its beginning point and hence eternal. No real change occurred on the moon or above, and, again, there were no empty spaces; the heavens were perhaps filled with an invisible celestial "quintessence" or fifth element, or perhaps the planets and stars were embedded in crystalline (invisible) spheres that fitted neatly together. Aristotle, like many other Greek thinkers, seems to have thought of the cosmos as all fitted together like a living creature, an organism, that was alive in its own self and that human beings could understand by rational and logical thought processes. [45; 55]

Nevertheless, even though Aristotle was still being taught in the universities, western Europe was by the seventeenth century saturated with machines and labor-saving devices that went far beyond the purview of Aristotelian patterns of thought: mechanical clocks that utilized the force of gravity; windmills and water mills that similarly exploited the motive forces of wind and water. The early mechanical genius of Newton's childhood constructions shows him aligned in significant ways with the mechanistic aspects of Western culture, so it comes as no surprise to find him adopting the new mechanical natural philosophies of the seventeenth century about halfway through his undergraduate career.

Mechanism and Other Influences on Newton

The appearance of mechanistic philosophies of nature in the seventeenth century was a broad-based phenomenon but one that was in sharp contrast to Aristotelianism, for the mechanical philosophies held that nature acts like a machine rather than like a living organism. No doubt encouraged by the prevalence of mechanical devices developed in Christian Europe, mechanical philosophies also had ancient roots in the works of Lucretius, a Roman poet of the first century B.C., and in the work of Epicurus, a Hellenistic Greek philosopher of the fourth and third centuries B.C. Revived by humanistic scholarship, interest in these ancient systems of thought, antithetical as they were to Aristotelian doctrines as well as Christian ones, had spread across national boundaries from Italy to northern Europe to England.

In general, the seventeenth-century philosophies that were at least partially based on ancient mechanism argued for the existence of very small—indeed imperceptible—particles that were all made of the same sort of matter and had only mathematical properties such as extension, size, shape, and perhaps weight. The particles could combine with each other to form larger material masses, or they could dissociate from each other and be re-formed in other ways. Such corporeal associations and dissociations accounted for much change in the natural world; the analogy frequently invoked was that of the alphabet: the letters can be associated in various ways to form words, and the words in turn dissociated to yield the basic letters that can then be re-formed into different words. But if the imperceptible particles of matter were thought to be like letters of the alphabet, similar to the movable type used in contemporary printing presses, the world as a whole was thought to be organized like an intricate machine designed and ordered by the Creator God of Judeo-Christian tradition. Just as human artisans designed and ordered the intricate mechanical clocks, printing presses, and wind and water mills that were common in western Europe, so the Deity had created a world-machine and set it in motion in ways that human beings could learn to understand. [19; 45]

Of the many varieties of mechanical philosophy available in the 1660s, those of the French philosophers René Descartes (1596–1650) and Pierre Gassendi (1592–1655) were important to Newton, as were those of the English philosophers Walter Charleton (1619–1707), Robert Boyle (1627–91), Thomas Hobbes (1588–1679), Kenelm Digby (1603–65), and Henry More (1614–87). Descartes had created the first total world system since that of Aristotle: his system was a plenum (completely full of matter) like that of Aristotle, but indefinite in extent rather than limited and bounded by the sphere of the fixed stars as Aristotle's had been. The matter in Descartes's system could be ground down into ever finer pieces. Gassendi, on the other hand, paid more attention to the ancient version of Epicurus, and so argued that the particles of matter were really atoms and therefore "uncuttable," which is what the Greek word *atomos* means. Uncuttable atoms moving in void (empty) space made Gassendi's system quite different from that of Descartes, and Gassendi's work was made widely known in England by an English version of it published by Charleton. Boyle had studied both Descartes's and Gassendi's systems but, declining to choose between them, utilized elements from each. Hobbes's version was much too materialistic in the eyes of most English natural philosophers, because Hobbes did not make sufficient room in his system for spirit, soul, and the Deity, and Henry More thought there were similar materialistic tendencies even in Descartes's version. Digby, who was not a very systematic thinker, was much influenced by Descartes but kept so many elements of Aristotelian thought also that his version of the mechanical philosophy was marked by severe inconsistencies. Although there indeed were many problems with these various mechanical philosophies, problems both scientific and theological, they were the cutting edge of natural philosophical thought in the seventeenth century, and after Newton absorbed their basic principles he seldom utilized Aristotelian doctrines again, though he was later to supplement mechanistic thought with other ancient systems in good humanistic fashion. [14; 17; 18; 27; 45; 47; 50; 51; 62; 79; 80; 81; 101; 104]

Newton's later fame could hardly have been predicted during his student years, but he was soon to tackle—and solve—many of the physical and mathematical questions that engaged his contemporaries. In January 1665 he took his Bachelor of Arts degree, but in the summer of 1665 he was forced to retire to his home at Woolsthorpe because the university was closed from an outbreak of bubonic plague (endemic in Europe since the first great pandemic in 1348–49 known as the Black Death). The university remained closed most of the time until the spring of 1667, and Newton's enforced sojourn at Woolsthorpe has come to be known as his *annus mirabilis*, the marvelous year in which he invented his "fluxions" (the calculus), discovered white light to be compounded of all the distinctly colored rays of the spectrum, and found a mathematical law of gravity, at least in a tentative form. [9; 40; 73; 101; 104]

The gradual development and unfolding to the world throughout subsequent years of the productions of that brief period were to establish his reputation upon the granite foundation it still enjoys. We will return to a consideration of each of these major achievements, which are still recognized and admired by the modern world, but we must also give due consideration to the fact that physical and mathematical problems were not the issues of greatest concern to most people in the seventeenth century, nor were they the issues of greatest concern to Newton himself. He lavished much more of his time on alchemy, church history, theology, prophecy, ancient philosophy, and "the chronology of ancient kingdoms."

Newton was born into the crucible of civil war in England, and the religious and political struggles of the period were to affect him deeply. The Reformation of the sixteenth century had shattered Christendom irrevocably, and, as religious sects had proliferated, especially in northern Europe, the intensity of political issues had grown as well, for church and state had formerly been perceived as a unity, like the two sides of a single coin. With such an understanding of the relationship of religious belief and political power, the notion of religious toleration was at first virtually inconceivable: divisions in creed and dogma were thus fought out on physical as well as intellectual battlefields. The full force of the tumult that had already torn Continental Europe for over a century arrived in England just at the time of Newton's birth. So even though Newton's long life carried him well into the eighteenth century and he came to be perceived as one of the principal founders of eighteenth-century Enlightenment thought, his own concerns remained centered to a great extent on the political and religious problems of the mid-seventeenth century, and he himself desired above all else to restore religion to the pristine purity, power, and centrality it had once enjoyed in human life. [6; 9; 18; 45; 50; 65; 66; 67; 99; 101]

In looking backward to the pure original religion he supposed humanity to have known and practiced at the beginning of time, Newton reflected the reverence for antiquity that had been the hallmark of Renaissance humanism. In many ways, indeed, Newton's intellectual development is best understood as a product of the late Renaissance, a time when the revival of antiquity had conditioned the thinkers of western Europe to look backward for Truth. The Renaissance humanists of the fourteenth century had rediscovered the glories of Roman poetry and prose, while in the fifteenth century a newly restored proficiency in the Greek language had revealed the awesome philosophical power and beauty of the works of Plato (427?–347 B.C.) and the Neoplatonists, as well as the mysterious doctrines of the Hermetic Corpus. Hermes Trismegistus (his surname meant "Thrice Greatest"), the supposed author of the Hermetic Corpus, was not a real person, although the thinkers of the Renaissance believed him to be not only real but also very ancient, perhaps even more ancient than Moses (who had composed several of the first books of the Judeo-Christian Bible). His supposed antiquity gave Hermes great authority in the eyes of Renaissance scholars, and Hermetic doctrines supported all sorts of magical, astrological, and alchemical enterprises in the sixteenth and seventeenth centuries. [18; 21; 52; 58; 69; 94; 99]

Humanist scholars revived innumerable other treatises from antiquity as well: treatises on medicine, mathematics, natural philosophy, astronomy, magic, alchemy, astrology, and cosmology, including the works of Lucretius and Epicurus mentioned above. Such materials were previously unknown in western Europe or known before only through inadequate translations. What an intellectual ferment they created! And as the new printing presses of western Europe spread the new editions of ancient works, and as the Renaissance spread north from its original Italian base, materials from Jewish, Egyptian, and Christian antiquity were added to the heady mixture, where they contributed to Reformation scholarship, to an interest in the Hebrew language and the cabala, and to early attempts to decipher Egyptian hieroglyphics. [5; 6; 21; 65; 69; 99]

Newton's Way of Thinking and Working

Thanks to this great revival of ancient thought, to humanist scholarship, to the quarrels of the Reformation, and to new developments in medicine, science, mathematics, and natural philosophy prior to or contemporary with his most intense period of study, Newton clearly had access to an unusually large number of systems of thought. Each system had its own set of guiding assumptions, so in that particular historical milieu some comparative judgment between and among competing systems was perhaps inevitable. One could hardly accept them all as equally valid. But such judgments were difficult to make without a culturally conditioned consensus on standards of evaluation. By the mid-seventeenth century the old verities in both religion and natural philosophy had been subverted but no new ones agreed upon.

A standard of evaluation was precisely what was lacking, a situation that led many people to adopt a skeptical attitude and to doubt that any true knowledge about the world or God could ever be attained. The formalized skepticism of Pyrrhonism had been revived along with other aspects of antiquity, but one may trace an increase in a less formal but rather generalized skepticism at least from the beginning of the sixteenth century, as competing systems laid claim to Truth and denied the claims of their rivals. As a consequence, western Europe underwent something of an intellectual crisis in the sixteenth and seventeenth centuries. What, indeed, was it possible for one to know without lingering and bewildering doubts? Among so many competing systems, how was one to achieve certainty? Could the human being attain Truth? [18; 69; 83; 84; 99]

Newton was not a skeptic. On the contrary, he seems to have adopted a contemporary response to questions of valid knowledge called the doctrine of "the unity of Truth," a position that was in fact one answer to the problem of skepticism. Not only did Newton respect the idea that Truth was accessible to the human mind, but also he was very much inclined to accord to several systems of thought the right to claim access to some aspect of the Truth. For those who adopted this point of view, the many different systems they encountered tended to appear complementary rather than competitive. The assumption they made was that Truth did indeed exist somewhere beyond the apparently conflicting representations of it currently available. True knowledge was unitary, and its unity was guaranteed by the unity of the Deity, He being the source of all Truth. As a practical matter, those who followed this doctrine of the unity of Truth became quite eclectic, which is to say that each thinker selected parts of different systems and welded them into a new synthetic whole that seemed to him (or her) to be closer to Truth. That was certainly Newton's method, and in the course of his long life he marshaled the evidence from every source of knowledge available to him: mathematics, experiment, observation, reason, the divine revelations in biblical texts, historical records, mythology, contemporary scientific texts, the tattered remnants of ancient philosophical wisdom, and the literature and practice of alchemy. [18; 84; 85]

One must realize, however, that in making selections from the various sources of knowledge available to him Newton utilized a sophisticated balancing procedure that enabled him to make critical judgments about the relative validity of each. Perhaps the most important element in Newton's contribution to scientific method as it developed in subsequent centuries was the element of balance, for no *single* approach to knowledge ever proved to be effective in settling the knowledge crisis of the Renaissance and early modern periods. Human senses are subject to error; so is human reason. So is the interpretation of revelation; so is the mathematico-deductive scientific method put forward by Descartes earlier in the century. Since every single approach to knowledge was subject to error, a more certain knowledge was to be obtained by utilizing each approach to correct the other: the senses to be rectified by reason, reason to be rectified by revelation, and so forth. [18]

The self-correcting character of Newton's procedure constitutes the superiority of Newton's method over that of earlier natural philosophers, for others had certainly used the separate elements of inductive reasoning, deductive reasoning, mathematics, experiment, and observation before him, and often in some combination. But Newton's method was not limited to the balancing of those approaches to knowledge that still constitute the elements of modern scientific methodology, nor has one any reason to assume that he would deliberately have limited himself to those familiar approaches even if he had been prescient enough to realize that those were all the future would consider important. Newton's goal was much broader than the goal of modern science. Modern science focuses on a knowledge of nature and only on that. In contrast, Newton's goal was a Truth that encompassed natural principles but also divine ones as well. He had a deep religious concern to establish the relationship between God and His creation (nature), and so he constantly searched for the boundaries between God and nature where divine and natural principles met and fused. As a result, Newton's balancing procedure included also the knowledge he had garnered from theology, revelation, alchemy, history, and the wise ancients. [18]

Blinded by the brilliance of the laws of motion, the laws of optics, the calculus, the concept of universal gravitation, the rigorous experimentation, and the methodological success, subsequent generations have seldom wondered whether the discovery of the laws of nature was all Newton had in mind. Scholars have often missed the religious foundation of his quest and taken the stunningly successful by-products for his primary goal. But Newton wished to look through nature to see God, and it was not false modesty when in old age he said he had been only like a boy at the seashore picking up now and again a smoother pebble or a prettier shell than usual while the great ocean of Truth lay all undiscovered before him. [18]

Eighteenth-, nineteenth-, and early twentieth-century views of Newton were developed almost entirely on the basis of his principal published works: *Philosophiae naturalis principia mathematica* (1687, 1713, 1726); *Opticks: or, a Treatise of the Reflexions, Refractions, Inflexions and Colours of Light. Also Two Treatises of the species and Magnitude of curvilinear Figures* (1704, 1706, 1721, and the posthumous edition of 1730); *Arithmetica universalis* (1707); *De analysi* and *Methodus differentialis* (1711). The Newtonian worldview, developed almost wholly on the basis of his successes in mathematics and physical science, so subtly and deeply colored the outlook of succeeding generations that the fuller seventeenth-century context in which Newton's thought had developed was lost to view. Thus it became a curious anomaly—and one to be explained away—that Newton's studies in astronomy, optics, and mathematics only occupied a small portion of his time. In fact most of his great powers were poured out quite otherwise.

The fact might never have been recognized, however, except for the survival of great quantities of manuscripts in Newton's hand. When Newton died in 1727 without leaving a will, his possessions passed to his niece, Catherine Barton Conduitt, and afterward to her descendants. The papers were examined with a view to possible publication later in 1727, and a few were published shortly afterward, but many of them were marked "not fit to be printed," and almost all of them were put back in their boxes. In the nineteenth century the family offered them to Cambridge University. The university appointed a group of men, mostly eminent scientists of the period, to examine the papers, and they selected for retention those focused on mathematics and physical science. These now comprise the Portsmouth Collection, University Library, Cambridge. The rest were returned to the family as being of no interest to the university and so remained largely unknown until they were sold at auction in 1936. The auction scattered them all over the world; although a number of them are still in the hands of private collectors, most of them are now held by research libraries and so are available for study. It is from detailed studies of these manuscripts of Newton's that our new and historically more accurate portrait of him has emerged. [7; 8; 11; 18; 23; 40; 61; 63; 65; 66; 67; 73; 76; 78; 87; 91; 98; 100; 101; 102; 103; 104]

Once one grasps the immensity of Newton's goal, many otherwise inexplicable aspects of his career fall into place. Now it is no longer necessary to explain away his fierce interest in alchemy or his dogged attempts at the correct interpretation of biblical prophecies, as many earlier biographers and scientists tried to do. If Newton's purpose was to construct a unified system of God and nature, as indeed it was, then it becomes possible to see all of his various fields of study as potential contributors to his overarching goal. It also becomes possible from this point of view to recognize that Newton's belief in the unity of Truth contributed greatly to his remarkable scientific creativity. For in the course of his long search for Truth he constructed many different partial systems and changed from one to another in ways that have often appeared erratic and inconsistent to later scholars. One may now see, however, that the pattern of change resulted from his slow fusion and selective disentanglement of essentially antithetical systems: Neoplatonism, mechanical philosophy, Stoicism; chemistry, alchemy, atomism; biblical, patristic, and pagan religions. It was precisely where his many different lines of investigation met, where he tried to synthesize their discrepancies into a more fundamental unity, when he attempted to fit partial Truth to partial Truth, that he achieved his greatest insights.

Newton's Early Mechanism: Cohesion and Gravity

With this broad view of Newton's work in hand, one may now begin to explore the intricate and marvelous development of the views that ultimately brought him so much acclaim.

Sometime during his student years Newton began to study the mechanical philosophers, as noted above, and he became a "corpuscularian." In the seventeenth century the term "corpuscularian" referred to anyone who believed that matter was comprised of small material particles or "corpuscles," whether the particles were understood to be infinitely divisible or whether there was supposed to be a limit to divisibility as in atomism, *atom* simply meaning "uncuttable," as we saw above. Newton chose elements of matter theory from Descartes, Gassendi (via Charleton), Boyle, Hobbes, Digby, and More and left a record of his thoughts in his student notebook. He seems to have accepted the notion that matter had least parts or atoms that were not further divisible. [18; 62]

At first Newton, like other mechanical philosophers of his time, placed considerable faith in the existence of an all-pervasive material medium that served as an agent of change in the natural world. By postulating a subtle aether, a medium imperceptible to the senses but capable of transmitting effects by pressure and impact, mechanical philosophers had devised a convention that rid natural philosophy of incomprehensible occult influences acting at a distance (e.g., magnetic attraction and lunar effects). For Newton just such a mechanical aether, pervading and filling the whole world, became an unquestioned assumption. By it he explained gravity and, to a certain extent, the cohesion of particles of matter. But because of the general passivity of matter in the mechanical philosophy, certain problems arose for many contemporary philosophers regarding cohesion and life, and eventually, for Newton, regarding gravity also. [18; 62]

The question of cohesion—that is, the problem of what makes the tiny corpuscles stick together—had always plagued theories of discrete particles, atomism having been criticized even in antiquity on this point. The cohesion of living forms seems intuitively to be qualitatively different from anything that the random, mechanical motion of small particles of matter might produce. Nor does atomism explain even mechanical cohesion in nonliving materials very well (such as, for example, the regular patterns in crystals of salts or gemstones), for any explanation of such regularity seems to require unverifiable hypotheses about the geometric configurations of the atoms or else speculation about their quiescence (or rest) under certain circumstances. [18; 51; 62; 105]

In the various forms in which corpuscularianism was revived in the seventeenth century, the problems remained and variants of ancient answers were redeployed. Descartes, for example, held that an external pressure from surrounding subtle matter (the aether) just balanced the internal pressure of the coarser particles that constituted the cohesive body. Thus no special explanation for cohesion was required, he claimed: the parts cohered simply because they were at rest close to each other in an equilibrated system. Gassendi's atoms, on the other hand, stuck together through the interlacing of antlers or hooks and claws, much as the atoms of Lucretius had before them, in what one might call a sort of primitive Velcro system. Charleton found not only hooks and claws but also the pressure of neighboring atoms and the absence of disturbing atoms necessary to account for cohesion. Francis Bacon (1561–1626) introduced certain spirits or "pneumaticals" into his speculations. In a system reminiscent of the Stoics, who were ancient critics of atomism, Bacon concluded that gross matter must be associated with active, shaping, material spirits, the spirits being responsible for the forms and qualities of tangible bodies, producing organized shapes and effecting digestion, assimilation, and so forth.

For Newton during his student years, with his mechanical aether ready at hand as an explanatory device, a pressure mechanism seemed sufficient to explain cohesion. He did not think that the simple resting (quiescence) of the particles close to each other could account for cohesion, but he did think that the "crowding" or pressure of the aethereal matter that filled all space might account for it. He noted the occasional geometric approach of Descartes but did not himself develop it. Newton was later to offer a radically different explanation of cohesion, one based on alchemical and Stoic considerations, but not while he was still an undergraduate. [18; 62; 68]

At first, in the 1660s, Newton considered gravity to be a mechanical mode of action. What is gravity, and what causes it? Modern science has still not found all the answers to these fundamental questions, but we know from common everyday experience that something makes bodies like ripe apples fall to the ground when they are detached from their trees, and that something will surely make us do likewise if we lean too far out an open window. Newton began to ponder the problem of gravity about 1664, and his first mechanical theory was derivative and nonmathematical, influenced by the theories of Descartes, Digby, and Boyle. Though the precise form was Newton's own, his theory was a restatement of impact physics, a conventional, orthodox (at that time) variety of mechanical philosophy: in short, an aether theory of gravitation. Bodies descend to earth, he said, through the impulsion of fine material particles; it is a mechanical stream of aethereal matter causing gravity, just as a flowing stream of water will carry wood chips downstream. [18; 62]

No hint exists in Newton's earliest statement of what gravity is later to become for him: an active principle (not mechanical) directly or indirectly dependent upon the activity of the Deity, the Creator God Who had made the world-machine and Who kept it in motion. Newton seems never to have focused solely on the material part of the natural world, as modern scientists usually do, but he always remained conscious of the presence of the Deity. Even in his undergraduate student notebook there is a recognition of God's omnipresence, the literally "being present everywhere" of the Deity that is later to subsume universal gravity in Newton's system of the world. When bodies are in motion in a world full of the aether, Newton said in this early notebook, some of the matter has to be crowded out of the way, so the motion meets with resistance. But in a vacuum that would not be the case. Even though God is present in the vacuum, God is a spririt and penetrates all matter, Newton added. God's presence causes no resistance, however, just as if nothing were in the way. Newton was to repeat much later his conviction that God is present where there is no body, as well as present where body is also present. There, as in the student notebook, God penetrates all matter. But whereas later the omnipresence of God and His ability to penetrate matter have the utmost significance with respect to gravity, that was not the case in the student notebook, where gravity was caused by the mechanical motion of small particles of matter to which God's presence simply constituted no obstacle. [18; 41; 62]

Newton's Early Mathematics: The Binomial Theorem and the "Fluxions"

At about the same time, in 1664, Newton began seriously to study mathematics. His preuniversity training had probably been limited to the basic rules of arithmetic, an elementary knowledge of weights and measure, and simple accounting techniques. Then he bought a book at a fair; he later called it a book on astrology, but it might equally well have been a book on astronomy, for the two terms were often used interchangeably in the seventeenth century. He could not understand it, however, not then being acquainted with trigonometry, so he bought a book on trigonometry only to discover that he was deficient in the background to that topic, never having studied the great fundamental work on plane geometry from antiquity, the *Elements* of Euclid. So he began to read Euclid. At first he found the propositions so easy to understand that he wondered why anyone would bother to write demonstrations of them, but he soon found propositions that were not intuitively obvious to him and then studied them with greater care. [104]

Once he had come to appreciate the logical power of Euclid's demonstrations, Newton turned to more modern mathematicians, reading the *Clavis mathematicae* of William Oughtred (1575–1660) and the *Geometry* of Descartes, both of which at first gave him some difficulty. By degrees he mastered them and soon moved on to the mathematical miscellanies of Franz van Schooten (1615–60) and the *Arithmetica infinitorum* of John Wallis (1616–1703). A few additional works apparently completed Newton's self-directed apprenticeship, and he began to discover and formulate new theorems of his own. [2; 3; 20; 73; 101; 103; 104]

The first of these of lasting significance was his discovery of the general binomial expansion in the winter of 1664–65, inspired by his reading of Wallis's work. A binomial is a mathematical expression with two terms in it, such as an x and a y. If the terms are added together, or if one is subtracted from the other, as in the expressions "$x + y$" or

"$x - y$," the expression with the two terms in it is called a binomial. Suppose, then, that one wishes to multiply the term "$x + y$" by itself (or square it). The procedure may be expressed in mathematical notation as $(x + y)^2$, where the 2 is known as the power or exponent to which the binomial is to be raised. The exponent might be 2 or any higher positive number (integer). If the designated multiplication is actually carried out, one obtains an expansion of the binomial, which in this case would be $x^2 + 2xy + y^2$, where the coefficients of the three terms are 1, 2, 1. Mathematicians prior to Newton's time had discovered rules for finding the coefficients for other positive powers to which the binomial could be raised, and such general rules were of great value in obtaining binomial expansions, for they made it unnecessary actually to carry out the lengthy process of multiplication. But whereas binomial coefficients for such positive integral powers had been known for some time, Newton's method was much more general, for it allowed the use of negative or fractional exponents also. [2; 20; 73; 104]

When the exponent is neither a positive integer nor zero, some binomial expansions constitute series with infinitely many terms. Newton was able to demonstrate that when such binomial expansions form infinite series the results do not just yield approximations (as mathematicians had previously supposed) but are subject to general definite laws, just as the algebra of finite quantities is. Infinite series expansions soon came to play a central role in his development of the calculus. [20; 73; 104]

His work on the fluxional calculus began in the autumn of 1664, and by the spring of 1665 Newton had resolved his several approaches into a general procedure for differentiation. Inspired in this case by Descartes's *Geometry*, in which algebra and geometry were combined to yield analytic geometry, Newton had focused on problems of finding tangents to curves, as well as normals and the radius of curvature at a general point. Newton regarded the curve as the locus of a moving point in a suitable coordinate system, the point itself being the intersection of two moving lines, one vertical and the other horizontal. The vertical and horizontal components changed with the "flux" or flow of time, and the "fluxions" of the variables were their derivatives with respect to time, indefinitely small and ultimately vanishing increments of the variables. By October of 1666 Newton had also mastered a general method for the reverse procedure, integration, to compute the area under the curve. Although various limited procedures for finding areas and tangents had been in use by the early seventeenth century, Newton's methods yielded general and systematic techniques and demonstrated the inverse relation between area problems and tangent problems. For a young man not yet 24, and apparently completely self-taught in this area, that was quite a dramatic achievement, and he himself said much later that he was then in the prime of his life for invention. But of course there was more to come. [20; 33; 73; 103; 104]

As Newton sat in the garden at Woolsthorpe one day during his *annus mirabilis*, some apples fell to the ground close to him. The event prompted him to consider the power of gravity that had brought the apples down and to speculate that the power of gravity might extend as high as the moon and help keep the moon in its orbit around the earth. The story of the apples is so well known now that in a recent set of British postage stamps commemorating the life and work of Isaac Newton one stamp had no design on it but an apple, the symbol of Newton's solution to the problem of gravity. [21; 100]

But the exact significance of the episode in Newton's developing thought on the subject of gravity has been less well understood. For a long time it was supposed that his creative genius allowed Newton at that time to formulate the law of universal gravity, so there was much scholarly speculation regarding the 20-year "delay" before he published his new discovery. Now the predominant interpretation of the episode of the falling apples is quite different: at most Newton then worked through an inconclusive comparison of the fall of an apple with the fall of the moon that yielded an approximation of the so-called inverse-square relationship, that is, the mathematical law that indicates that the power of gravity diminishes as the square of the distance between two gravitating bodies increases. It was only much later that the full generality and universality of the law emerged, for it required from Newton some major conceptual revisions. [12; 18; 40; 71; 73; 77; 78; 102]

In 1666, when the apples fell and he was prompted to consider also the fall of the moon, Newton had accepted the doctrine of a mechanical aether as the cause of gravity, as noted above. In particular, he had accepted Descartes's version of the aether, a version in which the aether swirled around the earth in a vortex pattern, a sort of imperceptible whirlpool that carried the moon in its orbit around the earth and accounted for the fall of objects like apples to the surface of the earth. Other planets had their own vortices also in Descartes's system, and around the sun there was supposed to be a giant whirlpool that carried the planets, including earth, around the sun in their regular cycles. Newton held to that general belief until about 1684. Indeed, in 1666 the manuscript evidence shows him thinking only about gravity with respect to the earth and the earth-moon system; he apparently did not generalize the system even to include the sun and other planets until 1675. So it seems that in 1666 his concept of gravity was far from the universal one that appeared in the *Principia* in 1687. [12; 18]

Acting against the downward pressure of the gravitational aether in Newton's early mechanics was the opposing endeavor to recede from the center, a centrifugal (center-fleeing) force. The term "centrifugal" was coined by Christiaan Huygens (1629–95), a Dutch natural philosopher and mathematician who was also a follower of Descartes. The centrifugal force accounted for the tendency of objects in circular motion to fly away from the center of motion, as in the case of a stone whirled at the end of a sling. Treated by physicists now as an illusory force, in the 1660s, 1670s, and early 1680s Newton accepted it as a real force that balanced and was balanced by the pressure of the vortex, the balance of the two forces being what kept the moon in its orbit. The same sort of analysis of forces appeared also in his early work on terrestrial mechanics during this period. But in the 1680s, when writing the *Principia*, Newton recognized the illusory nature of the centrifugal force and of the aethereal vortex as well. At that time he changed his analytic framework to two other opposing forces: inertia (the tendency of a body to continue in straight-line motion or at rest unless a force is applied to it) and a centripetal (center-seeking) force, a term he coined himself and one that reflects his new understanding of the center-seeking force as the mirror image of Huygens's center-fleeing force. We will return later to the insights of the 1680s that allowed Newton to revise his understanding of celestial mechanics, to generalize and universalize the force of gravity, and to construct a system of the world that would last virtually unchallenged for three centuries. [12; 18; 40; 71; 73; 77; 78; 102]

Newton's Early Optics: Particles and Prisms

Still in a white heat of creativity, what he later called his prime time of invention, Newton also began about 1666 to revise and reform contemporary views on the nature of light and color. Descartes had argued that the light from the sun that we see is simply a pressure in the aether. On the analogy of a blind person's walking stick, the pressure on the end of which is transmitted instantaneously to the hand, the sun's light sets up a chain of pressure from sun to eye that one then experiences as the sensation of light. That theory really would not do at all, Newton observed, because if light is just pressure on the eyeball, then one should be able to see perfectly well at night by running forward, since the forward motion of the runner would generate pressure of air and aether against the eye. [62]

Newton opted for a particulate definition of light, that is, the emission by the sun and other luminous bodies of extrafine corpuscles that we experience as light. Newton's theory of particulate emission implies that the transmission of light is not instantaneous but, on the contrary, must have a definite velocity. Early measurements of the speed of light later in the seventeenth century seemed to justify Newton's views, and light was then treated as particulate until the early nineteenth century. At that time a wave theory of light came to dominate scientific thought on the subject. Newton had recognized certain wavelike phenomena associated with light but had supposed them to be explained by motions in the all-pervading aether as the light corpuscles passed through it; he had denied that light itself might be a wave. It did not appear to him to spread out into the shadow of obstacles it passed (as water waves, for example, will do). More precise measurements in the nineteenth century having demonstrated conclusively that light does have the wave properties Newton had denied, his theory was thoroughly rejected for about a century—one of the very few cases of a total rejection of Newton's theories by later scientists. But modern quantum theory has at least partially reinstated Newton's views in the twentieth century, for it is now recognized that light does sometimes appear as quanta (tiny particulate packets of energy) while under other circumstances its continuous wave properties predominate. [75; 101]

As for colors, Newton tackled that issue with prisms. Robert Hooke (1635–1702) had published his *Micrographia* in 1665, a book rightly famous for its wonderful illustrations of observations made with the then new microscope. Hooke was an English natural philosopher, sometime employed as an experimental operator by Robert Boyle and later Curator of Experiments for the Royal Society of London for Improving of Natural Knowledge. In his book *Micrographia* Hooke argued for a theory of colors as oblique and confused pulses of light. The color blue was for Hooke such a pulse in which the weakest part comes first with the strongest following; the color red had its parts in reverse order. In general, according to Hooke, the colors formed a scale of strength between lightness and darkness, with red closest to pure white light and blue the last step before darkness. Newton attacked Hooke's theory at the root by analyzing white light into the spectrum of colors with the prism and showing how the different colors were refracted at different angles thereby. Refraction is the change of direction of a ray (of light, in this case) when it passes from one medium into another, from air into and out of the glass prism in Newton's experiment.

If the ray of white light passes into the glass at an oblique angle, it is split into all the colors of the rainbow, because each color acts as an independent ray and has its own precise and specific angle of changed direction. An oblong rainbow of separate colors thus becomes visible as the rays leave the prism. The colored rays kept their unique angles of refraction when passed through a second prism, Newton demonstrated, and they also could be recombined to constitute white light. So, Newton argued, the colored rays are the fundamental individuals, and white light is a confused mixture of them. With respect to the colors of bodies, he continued, it is clear that most of the colored rays are absorbed by the body while one is reflected back to our eyes, such as green, for example, in the case of most living plants. [75; 101]

Utilizing both experimental and mathematical analyses, Newton developed his insights into light and color over several years. When he was appointed to the Lucasian Chair of Mathematics at Cambridge in 1669 he gave his first set of lectures on his optical discoveries. He sent papers on optics to the Royal Society in the 1670s also, but the strong opposition his papers encountered from Hooke and others discouraged him from further publication to such an extent that he never published the full text of *Opticks* until 1704, by which time Hooke had died and so could no longer attack Newton's views. The laws of optics reported there were, however, virtually all established while Newton was less than 30 years of age. [71; 75; 101]

But light had significance for Newton that went far beyond the laws of optics, for in both Christian and Neoplatonic traditions light carried with it the aura of divinity. Light was God's first created creature in the Genesis account of creation, and it was both symbol and agent of divinity in Neoplatonism. The metaphysics of light had a long and distinguished career in Christian Europe; as Newton was soon to discover, it was also closely associated with divine creativity in the literature of alchemy. [15; 18; 42; 56; 63]

Newton's Early Alchemy:
Life, Cohesion, and Divine Guidance

Newton turned to a study of alchemy about 1668. It is possible that he had already learned some of the rudiments of chemistry before he entered Cambridge, when he lodged with a local apothecary (or druggist) in Grantham while attending the King's School there. But Newton certainly made himself the master of contemporary chemistry shortly after his return to Cambridge after the plague years, in 1667 or 1668, the date of the chemical dictionary he compiled then. Newton's turn to alchemy, however, was altogether a different enterprise from learning basic chemistry. [17]

Even though in the seventeenth century there was some overlap between chemistry and alchemy, and of course both fields shared some mutual interest in the manipulation and transformation of the different forms of matter by chemical techniques, chemistry and alchemy had quite distinct goals. Alchemy never was, and never was intended to be, solely a study of matter for its own sake. Nor was it, strictly speaking, a branch of natural philosophy, for there was a spiritual dimension to alchemy—a search for spiritual perfection for the alchemist himself or herself, or a search for an agent of perfection (the "philosopher's stone") that could transform base metals into silver or gold or perhaps could even redeem the world. It was in fact the spiritual dimension to alchemy that led Newton to study it, but his goal was not exactly one of the traditional ones. He perceived alchemy as an arena in which natural and divine principles met and fused, and he understood that through alchemy it might be possible for him to correct the theological and scientific problems of the seventeenth-century mechanical philosophies. [15; 16; 17; 18; 58]

Since sometime earlier in the 1660s Newton had been troubled by a theological problem, and he hoped alchemy could provide a solution. He was, as were his older contemporaries Isaac Barrow (1630–77), Henry More, and Ralph Cudworth (1617–88), alarmed at the atheistic potentialities of the revived corpuscularianism of their century, particularly of Cartesianism (the mechanical philosophy of Descartes). Although the ancient atomists had not really been atheists in any precise modern sense, they had frequently been so labeled because their atoms in random mechanical motion received no guidance from the gods. Descartes, Gassendi, and Charleton had been at pains to allay the fear that the revived corpuscular philosophy would carry the stigma of atheism adhering to ancient atomism. They had solved the problem, they thought, by having God endow the particles of matter with motion at the moment of creation. All that resulted then was due not to random corpuscular action but to the initial intention of the Deity. [6; 17; 18; 30; 79; 80; 81]

Later writers, going further, had carefully instated a Christian Providence among the atoms (where the ancients, of course, had never had it). Only Providence could account for the obviously designed concatenations (or organization) of the particles, and so, via Christianity, a fundamental Stoic critique of the ancient atomists actually came to be incorporated into seventeenth-century atomism. This development was all to the good in the eyes of most Christian philosophers:

THEATRUM CHEMICUM
BRITANNICUM.
CONTAINING
Severall Poeticall Pieces of our Famous
English Philosophers, who have written
the Hermetique Mysteries in their owne
Ancient Language.
Faithfully Collected into one Volume,
with Annotations thereon,
By ELIAS ASHMOLE, Esq.
Qui est Mercuriophilus Anglicus.

THE FIRST PART.

LONDON,
Printed by J. Grismond for NATH: BROOKE, at the
Angel in Cornhill. MDCLII.

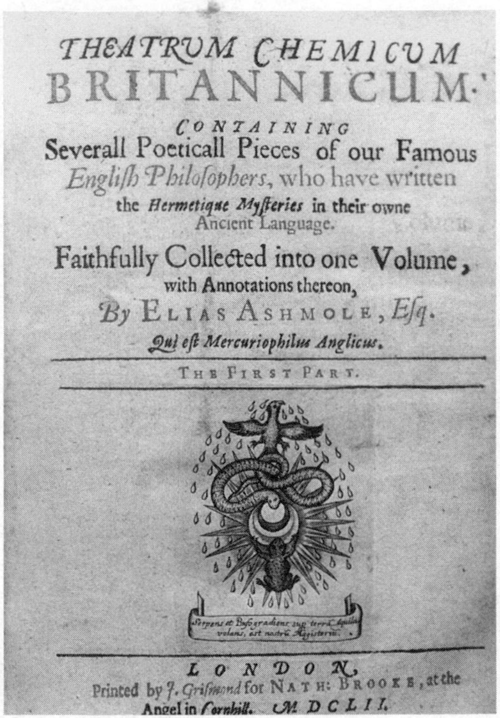

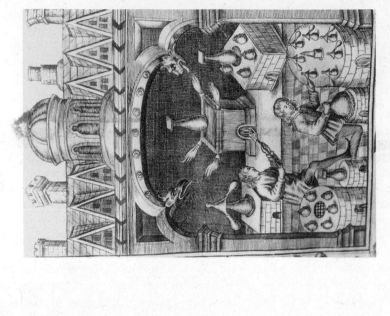

FIGURE 1.1 Title page of Newton's copy of Elias Ashmole's *Theatrum Chemicum Britannicum,* a collection of alchemical poetry that Newton studied with great care. Courtesy of the University of Pennsylvania Library, Rare Book Collection.

FIGURE 1.2 Illustration of seventeenth-century alchemists at work in their laboratory, from Ashmole's *Theatrum Chemicum Britannicum.* Courtesy of the University of Pennsylvania Library, Rare Book Collection.

atomism now supported religion, because without the providential action of God the atoms could never have assumed the lovely forms of plants and animals so perfectly fitted to their habitats. This was called "the argument from design" for the existence of a Deity: the presence of design (planned organization) in the natural world implied the existence of a Designer, a Deity Who had done the planning and organizing of the so obviously designed creatures in the natural world. The "argument from design" had quite ancient roots and had been present in Christianity from a very early period, but it assumed unparalleled importance in the seventeenth century. For if the new heliocentric astronomy, having moved the earth out of its central location in the cosmos, raised doubts about the focus of Providence upon such an obscure minor planet as earth now seemed, the new atomism seemed to relieve such doubts and reassure human beings that indeed divine Providence still cared for the world. [18; 57; 64; 96; 97]

The difficulty came when one began to wonder *how* Providence operated in the law-bound universe that was emerging from the new science, and that difficulty was especially severe in the Cartesian system, where only matter and motion were acceptable explanations. Even though Descartes had argued that God constantly and actively supported the universe with His will, it seemed to Henry More and others that Descartes's Deity was in danger of becoming a sort of absentee landlord, a Deity Who had set matter in motion in the beginning but Who then had no way of exercising His providential care. [18]

Newton faced this theological difficulty squarely and directly. The mechanical action of matter in motion was not enough. Granted that such mechanical action existed among the particles and could account for large classes of phenomena, yet it could not account for all. It could not account for the processes of life, where cohesive and guiding principles were clearly operative. It could not account even for the manifold riches of the phenomenal world. All forms of matter, never mind how various they appeared, could be reduced back to a common primordial matter according to the mechanical philosophers, but how had they been produced in the first place? From the particles of a universal matter with only primary mathematical properties, there seemed no sufficient reason for the forms and qualities of the phenomenal world to emerge at all. But emerge they did, and in such incredible and well-crafted plenitude that causal explanations based on mechanical interactions seemed totally insufficient. As Newton was finally to say in the General Scholium to the *Principia*, mechanical action (what he called "blind metaphysical necessity") could not produce variety because it is always and everywhere the same. Variety requires some further cause, a divine agent, and that is what he began to search for in his alchemical studies. [18; 30; 77]

In addition to learning contemporary chemistry and beginning his study of alchemy when he returned to Cambridge, Newton proceeded to Master of Arts and was elected a Fellow of Trinity College, a position that assured him an income and position in academic life—assured, that is, if he complied with the Fellowship regulation that he become an ordained Anglican priest within the next seven years. As we shall see, the prospect of ordination in the Church of England eventually became a source of deep anxiety to him.

Newton also polished some of his mathematical work from the Woolsthorpe period and showed portions of it to Isaac Barrow, who was then Lucasian Professor of Mathematics at Cambridge; he immediately put Newton's work into circulation (in manuscript form) among interested English mathematicians. The mathematical professorship had recently been created and endowed by Henry Lucas, and Barrow was its first incumbent. But Barrow soon resigned to return to his preferred life of theological study and preaching, and Newton was elected to replace him in 1669. His age was not yet quite 27 years. Already he could add the luster of his new optics and his new calculus to the Chair; in 1687 he would also add the *Principia*.

But the truth of the matter was that, for all the honor the Lucasian Professorship brought to Newton and for all the honor he brought to it, he had already immersed himself in the study of alchemy, and that was to be his most consuming passion for many years. Probably he had begun to read alchemical literature in 1668; in 1669 he purchased chemicals, chemical glassware, materials for furnaces, and the six massive folio volumes of *Theatrum chemicum*, a compilation of alchemical treatises. He established a laboratory of his own at Trinity College, and the records of his subsequent laboratory experimentation still exist in manuscript. Each brief, and often cryptic, laboratory report hides behind itself untold hours with hand-built furnaces of brick, with crucibles, with mortar and pestle, with the apparatus of distillation, and with charcoal fires; experimental sequences sometimes ran for weeks, months, or even years. He combed the literature of alchemy also, compiling voluminous notes and even transcribing entire treatises in his own hand. Eventually he drafted treatises of his own, filled with references to the older literature. The manuscript legacy of his scholarly endeavor is very large and represents a huge commitment of his time, but to it one must add the record of that extensive experimentation, a record that involves an amount of time impossible to estimate but surely equally huge. He seems to have continued his serious work on alchemy from about 1668 until 1696, when he left Cambridge for London and the Royal Mint, and even after 1696 he continued to study alchemical texts and to rework his own alchemical papers. [17; 18]

The focus of Newton's work in alchemy was already apparent in one of his very earliest independent alchemical papers; easily distinguishable from his reading notes and transcriptions, it is a short paper of alchemical propositions, in which he argued for the existence of a vital agent diffused through all things. This paper was probably written in 1669 (though Newton left it undated, as he did most of his papers), and it represents one of his earliest attempts to order the chaotic alchemical literature he was encountering.

The vital agent Newton described in that paper was universal in its operations. It had a general method of operating in all things but accommodated itself to the particular nature of particular subjects, and it assumed the particular form of each subject so as to be indistinguishable from the subject. In this manuscript Newton called the vital

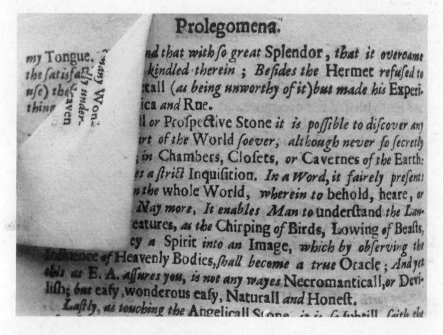

FIGURE 1.3   Newton folded down (or up) the corners of pages in his books to mark points of special interest to him. This fold down directs attention to a specific part of Ashmole's alchemical Prolegomena (or introduction) in Newton's copy of *Theatrum Chemicum Britannicum*. Courtesy of the University of Pennsylvania Library, Rare Book Collection.

agent "the mercurial spirit": later he was to call it a "fermental virtue" or the "vegetable spirit" and eventually, in the *Opticks*, the "force of fermentation." It was responsible for organizing particles of matter into all the various forms of the phenomenal world; it was also responsible for disorganizing them, for reducing organized forms back to the primordial particles. It was the natural agent God used to organize matter and put His will into effect in the natural world. [18]

Alchemy and the mechanical philosophies of the seventeenth century seem to have shared the doctrine of the unity of matter, the idea that ultimately all forms of matter were capable of being reduced back to a primordial condition in which all matter was alike and without form. In the philosophy of Aristotle, upon which much alchemical thought was predicated, such formless material would have been called "prime matter" and would not have been treated as particulate, as it was in the mechanical philosophies. Nevertheless, the notion that the material substances of ordinary, everyday experience could all be reduced to something more primal, something without the ordinary properties of color, taste, odor, texture, and so forth, was not foreign to either sort of matter theory. So there was nothing antimechanistic about the idea Newton expressed in his alchemical propositions paper: that something might act upon a substance to break down the formed aggregate and reduce it to a chaotic condition in which it had no ordinary properties.

On the other hand, it seems impossible to find a mechanistic counterpart to the agent itself, for it was indeed profoundly antimechanistic. It did not act by pressure or impact, as mechanistic particles did, but instead acted in a way that suggested design and willed or planned activity, for example, in the beautifully regular patterns that form in mineral crystallization, or in the changes that occur in fermentation as grape juice is transformed into wine, or in the marvelous transmutation of an acorn into an oak or an egg into a chick. There was no distinction made in seventeenth-century alchemical thought between what we would call the chemical and the biological realms. [15; 18]

Newton had become preoccupied with a process of disorganization and reorganization by which developed species of matter might be radically reduced, revivified, and led to generate new forms. The alchemical agent was able to cause death and putrefaction, returning matter to an unformed condition; but it was equally able to infuse the unformed matter with new life and to lead it to new forms of organization. For as he himself said later, all matter duly formed is attended by signs of life. The implication of that statement of Newton's is, of course, that matter in a formed condition is quite different from the passive ultimate particles of unformed matter. Formed matter has somehow acquired the quality of being alive, a quality conveyed to it by the active, vital alchemical agent that acts in the formation of everything. [18; 30]

From what sources has Newton derived his ideas on the universal vital agent that he is here busily attaching to seventeenth-century mechanism? Quite possibly only from alchemy at this early stage in his development, though his vitalistic ideas were soon reinforced by other sources, especially the Stoic philosophers. [18]

Vitalism seems to belong to the very origins of alchemy. In the early Christian centuries, when alchemical ideas were taking shape, metals had not been well characterized as distinct species. They were sometimes thought to have variable properties, like modern alloys. More frequently, they were thought to be like a mix of dough, in which the introduction of a leaven might produce desired changes by a process of fermentation, or even similar to a material matrix of unformed matter, in which the injection of an active male sperm or seed might lead to a process of generation. By analogy, alchemists referred to this critical phase of the alchemical process as fermentation or generation, and the search for the vital metallic ferment or seed became a fundamental part of their quest. Similar ideas occur in Aristotle and were commonplace in Newton's time.

Inspired by his interest in a vital agent, Newton had begun to grope his way toward mending the deficiencies of ancient atomism and contemporary corpuscularianism. He had concerned himself with life and cohesion. He now sought the source of all the apparently spontaneous processes of fermentation, putrefaction, generation, and vegetation— that is, everything associated with normal life and growth, such as digestion and assimilation, *vegetation* being originally from the Latin *vegetare*, "to animate, enliven." These processes produced the endless variety of living forms and could not be relegated to the mechanical actions of gross corpuscles, a point he emphasized in the 1670s and to which we will return later. Mechanical action could never account for the process of assimilation, in which foodstuffs were turned into the bodies of animals, vegetables, and minerals. Nor could it account for the sheer variety of forms in this world, all of which had somehow sprung from the common matter. [18]

Newton's Discovery of Stoic Philosophy and His Later Alchemy

The most comprehensive answer to such problems of life and cohesion in antiquity had been given by the Stoics. The Stoics postulated a continuous material medium, the tension and activity of which molded the cosmos into a living whole and the various parts of the cosmic animal into coherent bodies as well. Compounded of air and a creative fire, this medium was the Stoic *pneuma* (a Greek work meaning "breath" or "an airy matter") and was related to the concept of the "breath of life" that escapes from a living body at the time of death and allows the formerly coherent body in which it had resided to disintegrate into its disparate parts. Although always material, the *pneuma* becomes finer and more active as one ascends the scale of being, and the (more corporeal) air descreases as the (less corporeal) fire increases. The Stoic Deity, literally omnipresent in the universe, is the hottest, most tense and creative form of the cosmic *pneuma* or aether, pure fire or nearly so. The cosmos permeated and shaped by the *pneuma* is not only living, it is rational and orderly and under the benevolent, providential care of the Deity. Though the Stoics were determinists, their Deity was immanent and active in the cosmos, and one of their most telling arguments against the atomists was that the order, beauty, symmetry, and purpose to be seen in the world could never have come from random, mechanical action. Only a providential God could produce and maintain such lovely, meaningful forms, and this "argument from design" for the existence of a Deity was later adopted by Christian thinkers, as we saw above. The universe, as a living body, was born when the creative fire generated the four elements of earth, water, air, and fire; it lived out its life span, permeated by vital heat and breath, cycling back to final conflagration in the divine active principle, and always regenerated itself in a perpetual circle of life and death. [17; 32; 46; 54; 86; 88; 94; 96]

The original writings of the Stoics were mostly lost, but not before ideas of *pneuma* and *spiritus* (a Latin word with a similar meaning) came to pervade medical doctrine, alchemical theory, and indeed the general culture with form-giving spirits, souls, and vital principles, for Stoicism was one of the dominant philosophies of late antiquity. Spiritualized forms of the *pneuma* entered early Christian theology in discussions of the immanence and transcendence of God and of the Holy Ghost, just as the Stoic arguments that order and beauty demonstrate the existence of God and of Providence entered Christianity as the argument from design for the existence of a Deity. The creative emanations of Stoic fire melded with the creative emanations of light in Neoplatonism. In addition to this broad spectrum of at least vaguely Stoic ideas, excellent, though not always sympathetic, summaries of philosophical Stoicism were available in many of the learned authors of late antiquity: Cicero, Seneca, Plutarch, Diogenes Laertius, Sextus Empiricus, and others. By the seventeenth century ideas compatible with Stoicism were very widely diffused, and latter-day Stoics, Pythagoreans, Platonists, medical men, chemists, alchemists, and even the followers of Aristotle vied with each other in celebrating the occult (hidden, secret) virtues of a cosmic aether that was the vehicle of a pure, hidden, creative fire.

Nonetheless, such a vital aether or *pneuma* was to be found in its most developed form in philosophical Stoicism. It is probable, as Newton's concern for the processes of life and cohesion grew apace in the early 1670s, that he amplified his mechanical philosophy further by a close reading of the available literature on the Stoics. Virtually all of the scanty fragments of ancient Stoicism known today had already been recovered by western Europe during the Renaissance, and Newton had most of them. Newton could surely have reconstructed for himself a reasonably sophisticated and comprehensive knowledge of Stoic thought from books in his own library. [18; 34]

Such reading would have affected Newton's alchemy only in reinforcing certain critical ideas, for most of his early alchemical sources were distinctly Neoplatonic in tone, and in them the universal spirit or soul of the world already permeated the cosmos with its fermental virtue. But Stoic ideas would have affected his views on the mechanical aether of his student years. It seems one may conclude that if Newton had not read the Stoics, then he must independently have reached answers similar to theirs when confronted with similar problems, for by about 1672 the original mechanical aether of his student notebook had assumed a strongly Stoic cast.

Newton described his new vitalistic aether in an alchemical treatise of about 1672, "Of Natures obvious laws & processes in vegetation." There he described the earth as a great animal or vegetable that inhales an aethereal breath for its vital processes and exhales again with a grosser breath. He called the aethereal breath a subtle spirit, nature's universal agent, her secret fire, and the material soul of all matter. The similarity between this particular Newtonian aether and the Stoic *pneuma* is unmistakable: they are both material, and both somehow inspire the forms of bodies and give to bodies the continuity and coherence of form that is associated with life. Furthermore, Newton was quite explicit in this treatise that the processes of life, what he called vegetation, were similar in all three kingdoms of nature: the animal, the vegetable, and the mineral. In this treatise Newton made a sharp distinction between mechanism and the life processes of vegetation: "Nature's actions are either vegetable or purely mechanical," he said. As purely mechanical he listed two items of special interest, gravity (to which we will return later) and what he called vulgar or common chemistry. [15; 18:30]

In common chemistry, of course, nature (or the chemist) may effect many changes in textures, and so forth, but, Newton argued, that sort of change occurs just by rearranging the corpuscles. On the other hand, vegetative or growth processes require some further cause, and the difference between the two sorts of chemistry (mechanical and vegetable) is "vast and fundamental." [18:30] Vegetable chemistry in the mineral kingdom is what we usually call alchemy, for the alchemists believed that metals grow in the earth just as plants grow on the surface of the earth. Newton was convinced that metals were the only part of the mineral kingdom that vegetate, other mineral substances being formed mechanically. Vegetation in metals, of which the alchemists wrote, was thus the simplest case for study, the vegetation of the animals and vegetables in the other kingdoms of nature being obviously more complex. So in the vegetation of metals (alchemy) lay the most accessible key to the problem of nonmechanical action, the kind of divinely guided activity in nature that Newton thought was necessary to correct the overly mechanized system of Descartes. [15; 18]

Newton's distinction between mechanical and vegetable chemistry thus emerges as crucial to his solution of the theological problem posed by his Cartesian inheritance. Mechanical chemistry may be accounted for simply by the mechanical coalitions and separations of the particles and requires no further explanation. But for all that great class of beings that nature produces by vegetation, we must have recourse to some further cause. Ultimately the cause is God, and within the realm of vegetable chemistry one may find an area of continuing divine guidance of the world and of matter, an area of providential care. It is God's will that directs the motion of the particles of matter and guides them into their designed arrangements. The vital Stoic and alchemical agent, the subtle spirit of life, the secret fire in the earth's aethereal breath is thus simply the natural agent God uses in directing the motion of the passive particles of matter.

One may now see that Newton was concerned from the first in his alchemical work to find evidence for the existence of a vegetative principle operating in the natural world, a principle that he understood to be the secret, universal, animating spirit of which the alchemists spoke. His early conviction was amplified by Stoic doctrines on the breath of life, the Stoic *pneuma* or *spiritus* or vital aether. He later came to see analogies between the vegetable principle and light, drawing on the Christian and Neoplatonic metaphysics of light, and he also came to see analogies between the alchemical process and the work of the Deity at the time of the creation of the world, when matter was first guided into organized forms. But above all Newton thought, and continued to think for the rest of his life, that by the use of this active vegetative principle God constantly molded the universe to His providential design, producing all manner of generations, resurrections, fermentations, and vegetation.

In short, the action of the secret animating spirit of alchemy kept the universe from being the sort of closed mechanical system for which Descartes had argued. Left to run by itself without provision for divine providential care, the Cartesian universe threatened traditional Christian values. Newton thought that belief in the Cartesian mechanical system, where matter filled all space and there was little or no room for spirit, and where no divine guidance seemed to be required on a daily basis, could lead to a materialist philosophy, to deism, or even to atheism. [18; 50; 62]

A materialist would emphasize matter to the exclusion of spirit; a deist might still believe that a Deity created the world and set it in motion but then left it to run by itself; an atheist would deny the existence of the Deity completely. Newton had a horror of all those philosophico-religious positions and was determined to do everything he could to counteract them. He thought that an irrefutable scientific demonstration of divine providential guidance of the small particles of matter would provide the needed evidence for the existence and activity of the Deity and would restore to humanity the true religion that had been lost. Given the importance of what he hoped to gain from his alchemical studies, it is easy enough to understand why he experimented and studied in that field with such persistence, year after year after year. None of Newton's convictions in this area of his work ever suffered substantial change, and though later he revised his terms "vegetable" and "mechanical" to the terminology of "active" and "passive" (terms he had learned from Stoic philosophy), the new terminology served exactly the same metaphysical purpose as the old. The foundational thinking about "active" (divine) and "passive" (material) principles that Newton first developed in his alchemical work later supported his basic patterns of thought in both the *Principia* and the *Opticks*, but let us next see how his other religious studies developed. [18]

Newton's Work in Other Religious Subjects

We have just seen that Newton's work in alchemy had a religious motivation, for he was convinced that a demonstration of divine activity in the guidance of the passive particles of matter was possible through alchemy. Other areas of his interests were even more directly and obviously focused on religion, but the ultimate goal of one of them at least was virtually identical to his goal in studying alchemy. That was the correct interpretation of biblical prophecy and its correlation with the recorded events of history, for such a correlation would also demonstrate divine activity in the world. Newton began work on the prophecies in the 1670s if not earlier, and he is thought to have still been working on his last version of their interpretation the night before he died in 1727. [18; 24; 101]

As alchemy was the story of God's ongoing activity in the world of matter for Newton, so history was the story of God's ongoing activity in the moral world, and as such it was a key for the interpretation of prophecy. Prophecy in the Bible was divinely inspired, and Newton spent untold hours on the writings of Daniel and the Apocalypse of St. John. But human beings could fully understand prophecy only after it had been fulfilled, for it was written in "mystical" language that was not readily accessible. In any event, Newton argued, a person was not to presume to interpret it with an eye for concrete prediction of the future. Only after the prophesied events had occurred could one see that they had been the fulfillment of prophecy. Then God's action in the world was demonstrated. [18; 24; 67; 71; 74; 76; 101]

The folly of Interpreters has been, to foretel times and things by this Prophecy [John's], as if God had designed to make them prophets. By this rashness they have not only exposed themselves, but brought the Prophecy also into contempt. The design of God was much otherwise. He gave this and the Prophecies of the Old Testament, not to gratify men's curiosities by enabling them to foreknow things, but that after they were fulfilled they might be interpreted by the event, and his own Providence, not the Interpreters, be then manifested thereby to the world. For the event of things predicted many ages before, will then be a convincing argument that the world is governed by providence. [74:251]

Newton's methodology in prophetic interpretation was undoubtedly influenced by the methods of others, particularly of recent Protestant interpreters including Joseph Mede (1586–1638) and Henry More, both of Cambridge University. Yet there was in addition something peculiarly Newtonian about it. In Newton's mind history seemed to bear a direct correspondence with experimental or even mathematical demonstration. Just as an experiment might enable the investigator to decide between alternative theories of natural phenomena, so historical facts might enable the interpreter to choose between possible interpretations of prophecy. For Newton only the firm correspondence of fact with correctly interpreted prophecy provided an adequate demonstration of God's providential action. What had been adumbrated or prophesied by divine agency in the prophecy had then been fulfilled by divine agency. What God had said He would do, He had done. That, and only that, provided for Newton a "convincing argument" for God's providential governance of the moral world. When actual historical developments exactly matched predicted ones in "the event of things predicted many ages before," one hears an echo of that universally satisfying geometrical conclusion, *quod erat demonstrandum*: QED. [74:251–52] That was exactly what Newton wanted to demonstrate with his prophetic and historical studies—God's providential action in the moral world—just as he desired by his alchemical studies to demonstrate God's providential action in the natural world. [18; 24; 74; 99]

Probably Newton labored over prophetical interpretations for fifty years, if not more, but in another part of his religious work he labored intensively for only a few years. After that relatively brief time, he was convinced that the entire Christian tradition since the fourth century had been in error and that he, Newton, had come closer to the Truth of primitive Christianity. Afterwards he adhered to his new convictions in spite of the problems they caused him in his own society. The issue was a doctrinal or theological one having to do with the nature of the Deity. Orthodox Christian doctrine in the seventeenth century was trinitarian; that is to say, the accepted belief was that God was "Three-in-One" or "One-in-Three," one God in Three Persons (God the Father, God the Son, and God the Holy Ghost) all coequal and coeternal and ultimately One. Newton disagreed. [22; 23; 76; 101]

The problem arose initially because Newton's Fellowship at his Cambridge college required that he accept ordination as a member of the clergy of the Church of England after seven years, or else resign his Fellowship, as we saw above. During the years preceding that deadline, Newton understandably immersed himself in theological studies, and in so doing read exhaustively in the patristic literature, that is, the treatises written by the fathers of the church in the early centuries of Christianity. Among those documents he found traces of the views of Arius, a theologian of the third and fourth centuries, and of the debates on the nature of Christ (God the Son) that culminated in the decision of a church council, the Council of Nicaea in 325, that Christ, God's

Son, was of the same substance as God the Father and was "begotten, not made, being of one substance with the Father," in the words of the Nicaean Creed issued by the council. Arius, who lost the argument, had believed that the Son was created, not begotten, and was *not* of the same substance as the Father. Newton decided, on the basis of the documents in the case, that Arius was right and that all of Christendom had been in error since 325. In the eyes of his contemporaries, however, had they but known of Newton's decision for Arius, Newton would have been the heretic and in mortal danger of losing not only his college Fellowship but also his Lucasian Chair of Mathematics. [28; 82; 101]

The viewpoint of Arius that Newton accepted was not trinitarian and thus was not orthodox, for in Arian theology the Son was created by the Supreme Deity and so was not coeternal with the Almighty God. Newton did think for a while that he would lose his Fellowship, but he cautiously let it be known that he preferred not to be ordained even while maintaining a discreet silence on the reason for that preference. The outcome was favorable for Newton: a permanent dispensation from the Fellowship requirement for ordination was obtained from the crown for the Lucasian Professor. Thus Newton was saved: he neither had to perjure himself by claiming to believe the trinitarian doctrine that he no longer believed, nor did he lose his university and college positions. [101]

Newton was so discreet about the whole matter that hardly anyone knew until the twentieth century that Newton had become an Arian in the early 1670s. Religious heterodoxy is not always a burning issue in the modern world, but getting at the Truth was a burning issue for Newton, and he was clearly prepared to relinquish his academic honors for the sake of his convictions. He remained a convinced Arian to the end of his days, and late in life he formulated an Arian creed that he presumably hoped would replace the Nicaean Creed for all believers. His Arian theology had an interesting impact on his natural philosophy also, for, as we have seen, Newton believed that Truth from any area of his studies should coalesce with Truth from any other area. We will return to a discussion of Arianism in connection with Newton's changing ideas on the cause of gravity, and in Part II we will see that a number of Newton's followers also became Arians in theology. [18; 22]

Newton's Mechanistic Theories of Gravity

As we have seen, when Newton first learned about the new mechanical philosophies of nature in his undergraduate years, he adopted a mechanistic explanation of gravity, relying on a shower of imperceptible aethereal particles that pressed bodies down toward earth. In his student notebook he even sketched two bits of machinery that might be built to take advantage of the shower of fine particles to produce perpetual motion, one machine constructed like a water wheel, the other with vanes somewhat like those of a windmill. Each was designed to operate from the impacts of the mechanical stream of aethereal matter causing gravity. [62]

A similar but more developed mechanistic explanation of gravity appeared a few years later in another of Newton's private papers, the alchemical treatise "Of Natures obvious laws & processes in vegetation," probably written about 1672. That may seem a very odd place for Newton to offer a speculative scenario about gravity, but that treatise is one of the prime exemplars of a small group of papers in which Newton was trying to fit partial Truth to partial Truth from some of his different lines of investigation. So although the mechanical aether for gravity and the vitalistic aether that carried the secret vivifying fire of the alchemists had originated in quite different studies in his early work, in this alchemical treatise Newton had combined them. The complex aether Newton then described followed a great circulatory path, not unlike that of the Cartesian vortices. It swept down to earth, making bodies heavy, but its finer and more active parts also provided the vivifying alchemical spirit. When it reached the earth, it continued into the earth's interior where it helped to generate air. The air in turn ascended from the interior to constitute the atmosphere, vapors, clouds, and so forth, until it reached the aethereal regions above. There the air pressed on the aether, forcing it to descend again toward the earth. [15; 18]

In 1675 Newton sent a paper to the Royal Society in London that contained a very similar system, but one in which the gravitational and vegetative functions of the aether were even more thoroughly combined. The movement up of air and the movement down of aether continued, with the air being "attenuated into its first principle" of aether when it reached the great aethereal spaces above. "For nature is a perpetuall circulatory worker . . . , Some things to ascend & make the upper terrestrial juices, Rivers and the Atmosphere; & by consequence others to descend for a Requittal to the former." [18:103; 71]

But in the 1675 paper there is one striking difference, and that is that the whole speculative system has moved toward universality. Appearing almost as an afterthought at the end of his description of the earthbound circulatory pattern, the operations of the gravitational-vegetative aether expanded to include the solar system. [18; 71]

And as the Earth, so perhaps may the Sun imbibe this Spirit copiously to conserve his Shineing, & keep the Planets from recedeing further from him. And they that will, may also suppose, that this Spirit affords or carrys with it thither the solary fewell [fuel] & materiall Principle of Light; And that the vast aethereall Spaces between us, & the stars are for a sufficient repository for this food of the Sunn & Planets. [71]

Newton's speculative aethereal system was enormously expanded in its scale of application when he thus extended it to the sun and other planets, but in no way did that expansion affect the mode of operation of the system, for both gravitational and vegetative functions were still attributed to the "Spirit." The "Spirit" had the stated gravitational function of keeping the planets in their orbits: the sun imbibed the "Spirit" to "keep the Planets from recedeing further from him." The vegetative function of the "Spirit," on the other hand, is readily apparent in other phrases: this spirit provided "food" and "fewell." It furthermore carried with it the "materiall Principle of Light," a sharp and unmistakable echo of Newton's identification of the vegetable spirit with "the body of light" in an earlier alchemical paper. [15; 18:104]

This combination of functions, both gravitational and vegetative, in Newton's speculative aethers was not to last, however. By 1679, in a letter to Robert Boyle, Newton had completely separated them and had formulated two new mechanistic scenarios to explain gravity. The new systems did not rely on a stream of aethereal particles as the old ones had done. Instead of a stream of particles flowing like a stream of water, Newton used in 1679 a nonmoving aether that was more dense in some places than in others. In both systems described to Boyle, however, gravity was again fully mechanized and detached from the vital alchemical agent.

Newton told Boyle that one of the gravitational conjectures in his letter came into his mind only as he was writing the letter, and, though we will meet a variant of that particular aethereal system again, it did not have a very long life in its original form. For a few months later, toward the end of 1679, Robert Hooke's challenges to Newton regarding the motion of bodies set Newton on a course of development that changed forever the conditions for aethereal speculation. [18; 71]

The correspondence with Hooke provided the stimulus for Newton's first solution to the problem of celestial dynamics in the terms later to appear in the *Principia*. As we have seen, Newton's earlier analysis of celestial motion had been cast in terms of a center-fleeing (centrifugal) force counterbalanced by the pressure of the aethereal vortex that carried the moon around the earth or the planets around the sun. In

1679 Hooke argued for a different way of approaching the problem: an attractive force *toward* the center of the orbit (what Newton later called the centripetal or center-seeking force), counterbalanced by the tendency of the planet or moon to move away from its orbit in a tangent (a straight line only touching the orbit in one place), due to its inertia. Having been very busy with his alchemical and theological studies during the 1670s, Newton had barely considered celestial dynamics quantitatively and had done no original work in that area since the 1660s. Hooke diverted him from his other studies and irritated him by correcting some errors Newton had made and by what Newton called Hooke's "dogmaticalnes," so Newton was "inclined" to try Hooke's mode of analysis using the centripetal force and inertia and then found the theorem by which he "afterward examined ye Ellipsis." It is possible that Newton made his trial of the new method late in 1679 or early in 1680, but even if so he once more quickly put his calculations aside for other studies, primarily alchemical and theological ones. [18:119; 71; 101]

Newton's conceptualization of gravity remained unsettled for several years following his interchange with Hooke. In an exchange of letters with Thomas Burnet (1635–1715) late in 1680 and early in the following year, Newton suggested a mechanism of vortical pressure for gravity, discussed the centrifugal force of the planets (a component of his pre-Hookian dynamical analysis), and mentioned "gravitation towards a center" without offering any mechanism for it. When conferring with John Flamsteed (1646–1719), Astronomer-Royal of England, about the comet of 1680, Newton mentioned the "attraction of ye earth by its gravity" but also mentioned the "motion of a Vortex." He was willing to "allow an attractive power" in the sun "whereby the Planets are kept in their courses about him from going away in tangent lines," which seems to presuppose Hooke's analysis, but, in refuting Flamsteed's notion that such an attraction might be magnetic, Newton utilized both the idea of the sun's vortex and the concept of the centrifugal force. About 1682 he referred to the material fluid of the heavens that gyrates around the center of the cosmic system according to the course of the planets. Not until 1684 do Newton's papers reflect the clarity of thought on dynamical principles that enabled him to launch the writing of the *Principia*, and only in the course of writing that work did Newton confront the problems that inhered in all his various early aethereal gravitational systems. [18:126; 71; 101; 102]

The *Principia*: Composition and Content

Edmond Halley (1656–1752), Fellow of the Royal Society and later Astronomer-Royal, was a central behind-the-scenes figure in stimulating the writing of Newton's most important work and in seeing it through the press (editing it, correcting proof sheets, drawing geometric figures, and even funding the publication himself). *Philosophiae naturalis principia mathematica* (*The Mathematical Principles of Natural Philosophy*), published in London in 1687 and now usually designated simply by its abbreviated Latin title as *Principia*, was the capstone of the Scientific Revolution of the sixteenth and seventeenth centuries and is often said to be the greatest work of science ever published.

The occasion for Newton to write the book arose in the following manner. Halley, Robert Hooke, and Christopher Wren (1632–1723), the great architect responsible for rebuilding many of the churches of London after the Great Fire of 1666, met in London early in 1684 and discussed a problem in celestial mechanics associated with the sun-centered astronomy of Copernicus: what curve would the planets describe if the force of attraction toward the sun varied inversely with the square of the distance of the planets from the sun? The three men reached no firm conclusion, but Halley, having heard that the Lucasian Professor in Cambridge might have some ideas on the subject, traveled to Cambridge in August of 1684 and put the question to Newton. Newton immediately responded that the curve would be an ellipse. When Halley asked Newton how he knew that, Newton responded that he had calculated it. Newton could not find his papers on the subject but promised Halley to do the work again and send the papers to Halley in London. [73; 101]

Whatever Newton might have done before on this matter, it was nothing compared to the power and generality of the work he then produced. By November of 1684 Newton had written and sent to London the first fruits of his new work, a tract on the motion of bodies in orbit. In it he defined centripetal force for the first time: "that by which a body is impelled or attracted towards some point regarded as its centre." He also defined resistance: "that which is the property of a regularly impeding medium," but he hastened to add that for his first several propositions on celestial dynamics "the resistance is nil." Thus by November of 1684 Newton knew the motions of celestial objects were not impeded by the medium through which they moved. [18:130; 73; 101]

If, however, the heavens were filled with the hypothetical aether of the mechanical philosophers, that aether should constitute a resisting medium. Unless the medium is somehow disposed to move with exactly the same variable speed that the planetary body exhibits, the planet should encounter enough resistance from the medium to cause an observed deviation from the mathematical prediction, just as projectiles in the terrestrial atmosphere are observed to deviate from mathematical prediction. [18; 73; 78]

Newton's realization that no form of the hypothetical gravitational aether of the mechanical philosophy could be reconciled with actual celestial motions cleared the deck, so to speak, for the development of his mathematical law of universal gravity, but it must have been rather a shock to him at first. From the time of his introduction to mechanical philosophy in the 1660s until early in the 1680s, he left an extensive record of his aethereal speculations, as we have seen. Even as he modified his schemata from time to time, he seems never to have doubted that the cause of gravity was some sort of material aether. But if the heavens were filled with such an aether, then its presence should produce some notable retardation on the motions of bodies passing through it, and none was in evidence. Newton had had to rethink all his aethereal mechanisms in order to make that statement in the first draft of his small tract on the motion of bodies, and one result of his rethinking is already in evidence in his definition of centripetal force. Whereas in earlier documents Newton had offered explanations of apparent *attractions* in terms of aethereal *impulsions* (impacts or pressure that pushed rather than pulled or attracted in a mysterious fashion), in the new definition he equivocated. Bodies are "impelled or attracted" by a centripetal force, he said. No causal mechanism was suggested nor any preference indicated between the two ways of describing the action of the force, a stance he was soon to adopt in the *Principia* itself. [18:132]

Newton left a great many papers associated with the writing of the *Principia.* He wrote and revised many times as he worked his way through one major conceptual or mathematical or observational difficulty after another. He worked at a very high level of involvement and creativity, by all reports—forgetting to eat, sleeping little. If he started out to go somewhere, a new idea or a fresh solution to an old problem might strike him, in which case he would return to his desk immediately and begin to write again, often even forgetting to sit down. The final work, passing bit by bit through Halley's hands to the press, eventually took the form of three books: the first two were severely mathematical treatments of the motions of bodies and of terrestrial and celestial mechanics, and the last one applied his mathematical discoveries to explicate the system of the world in which we live. It was our own solar system Newton described: the planets (including our earth) move around the sun is elliptical orbits and lesser satellites orbit some of the planets, and the distant stars are understood to be suns similar to our own. The work laid out the grand design of the universe that was accepted by educated people everywhere until modified by Einstein in this century, and it established the basic parameters of classical physics and mechanics. [10; 11; 12; 21; 52; 73; 77; 101]

The problems Newton solved in the *Principia* had become more and more urgent with each stage of development in astronomy and natural philosophy since the sixteenth century. The astronomy of the ancients had been geocentric: that is, with a stable and unmoving earth stood in the center of the cosmic spheres, while the planets and the stars revolved around the earth at constant speeds in circular patterns. It was in the nature of heavenly bodies to move in circles, a perfect and eternal geometric form, the ancients had said; perhaps also, some speculated, the planets and stars were embedded in crystalline spheres that carried them around in their endless circuits, or perhaps they were moved by angels. When Nicholas Copernicus (1473–1543) replaced the central earth with the central sun in 1543, creating a heliocentric (sun-centered) system, the earth was set in motion and became just another planet in orbit around the sun. That was very difficult for people to accept: the earth does not feel to us as if it is in motion; also, traditional Aristotelian physics assumed an immobile earth at the center, so a new physics was required for a moving earth. [10; 52]

Galileo Galilei (1564–1642) supplied the physics for a moving earth, showing how two or more motions could be compounded together, which Aristotle had claimed was impossible. So if the earth moves, everything on the surface of the earth moves with it, as well as having individual motions. Thus a cannonball fired straight up will return to land in the cannon's mouth (wind and other minor irregularities neglected), for the cannonball continues to share the earth's motion even as it ascends and descends. Galileo also found a mathematical law for the acceleration of falling bodies: the distance traveled is proportional to the square of the time the body has been falling from rest, and the constant of proportionality is the acceleration of the body due to gravity. So those that understood it were satisfied that a new physics for a moving earth was possible. But, on the other hand, Galileo's work did not explain much about motions in the heavens. [4; 10; 55]

In the meantime Tycho Brahe (1546–1601), a Danish nobleman, had collected vast new quantities of astronomical data. Utilizing huge observational instruments of his own design, he had obtained the most accurate naked eye (pretelescopic) astronomical data then available,

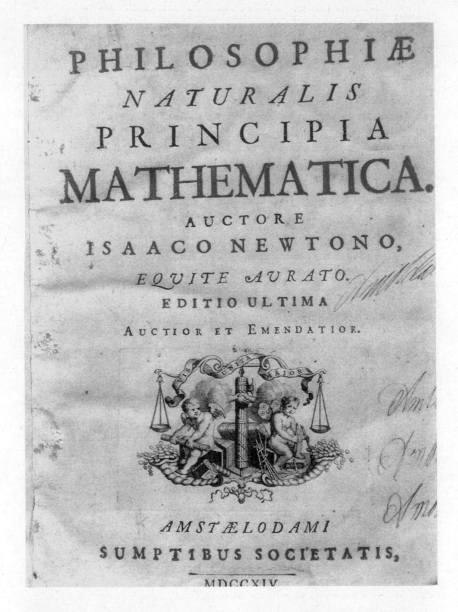

FIGURE 1.4 Title page of Newton's *Principia*, from the edition published in 1714 in Amsterdam. Interest in Newton's achievements in the *Principia* spread rapidly across Continental Europe. Courtesy of the University of Pennsylvania Library, Rare Book Collection.

PRINCIPIA MATHEMATICA. 3⟨

ter fe in ultima proportione Sinuum vel Tangentium anguloru
contactuum ad radios æquales pertinentium ubi radii illi in infin
tum diminuuntur. Attractio autem Lunæ in Terram in Syzygiis ⟨
excelfus gravitatis ipfius in Terram fupra vim Solarem 2 PK (⟨
de *Figur. pag.* 394.) qua gravitas acceleratrix Lunæ in Solem ⟨
perat gravitatem acceleratricem Terræ in Solem. In Quadrat
ris autem attractio illa eft fumma gravitatis Lunæ in Terram
vis Solaris KT, qua Luna in Terram trahitur. Et hæ attra
tiones, fi $\dfrac{AT+CT}{2}$ dicatur N , funt ut $\dfrac{178725}{ATq} - \dfrac{2000}{CT\times N}$

$\dfrac{178725}{CTq} \times \dfrac{1000}{AT\times N}$ quam proxime; feu ut $178725\ N\times CT$
$- 2000\ ATq\times CT$ & $178725\ N\times ATq + 1000\ CTq\times AT$. Na
fi gravitas acceleratrix Lunæ in Terram exponatur per numeru
178725, vis mediocris ML, quæ in Quadraturis eft PT vel T⟨
& Lunam trahit in Terram,
erit 1000, & vis mediocris
TM in Syzygiis erit 3000; de
qua, fi vis mediocris ML
fubducatur, manebit vis 2000
qua Luna in Syzygiis diftra-
hitur a Terra, quamque jam
ante nominavi 2 PK. Velo-
citas autem Lunæ in Syzygiis
A & B eft ad ipfius velocita-
tem in Quadraturis C & D,
ut CT ad AT & momentum
areæ quam Luna radio ad
Terram ducto defcribit in Syzygiis ad momentum ejufdem
areæ in Quadraturis conjunc-
tim; id eft, ut 11073 CT ad
10973 AT. Sumatur hæc ratio bis inverfe & ratio prior
femel directe, & fiet curvatura Orbis Lunaris in Syzygiis ad e⟨

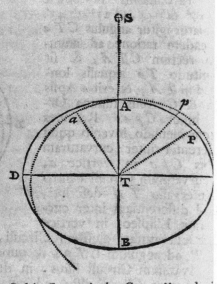

FIGURE 1.5 A passage from Book I of Newton's *Principia* on the orbit
of our moon. Since both the earth and the sun exert major gravitational
attractions on the moon, its orbit is quite irregular. Newton once said
the moon's motion gave him a headache. Courtesy of the University of
Pennsylvania Library, Rare Book Collection.

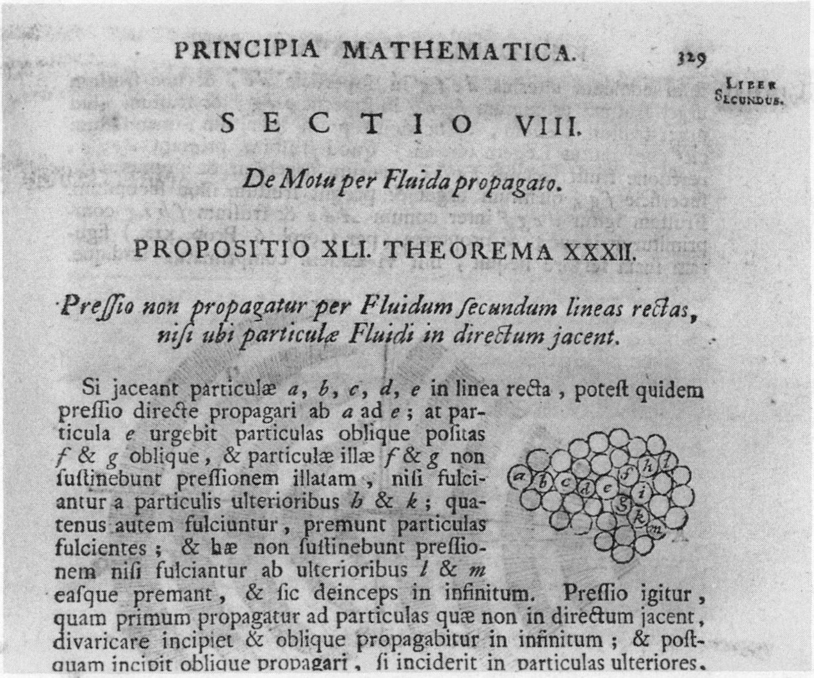

PRINCIPIA MATHEMATICA. 329

LIBER SECUNDUS.

SECTIO VIII.

De Motu per Fluida propagato.

PROPOSITIO XLI. THEOREMA XXXII.

Preſſio non propagatur per Fluidum ſecundum lineas rectas, niſi ubi particulæ Fluidi in directum jacent.

Si jaceant particulæ *a*, *b*, *c*, *d*, *e* in linea recta, poteſt quidem preſſio directe propagari ab *a* ad *e*; at particula *e* urgebit particulas oblique poſitas *f* & *g* oblique, & particulæ illæ *f* & *g* non ſuſtinebunt preſſionem illatam, niſi fulciantur a particulis ulterioribus *h* & *k*; quatenus autem fulciuntur, premunt particulas fulcientes; & hæ non ſuſtinebunt preſſionem niſi fulciantur ab ulterioribus *l* & *m* eaſque premant, & ſic deinceps in infinitum. Preſſio igitur, quam primum propagatur ad particulas quæ non in directum jacent, divaricare incipiet & oblique propagabitur in infinitum; & poſtquam incipit oblique propagari, ſi inciderit in particulas ulteriores,

FIGURE 1.6 A passage from Book II of Newton's *Principia* on fluid mechanics. Newton's propositions on fluids in Book II proved to be somewhat less successful than his explication of the system of the world. Courtesy of the University of Pennsylvania Library, Rare Book Collection.

and with Brahe's death in 1601 these data had passed to Johannes Kepler (1571–1630). Kepler, who was a Copernican and the Imperial Mathematician to Emperor Rudolf II in Prague, spent years with the data for Mars, calculating and recalculating in an effort to find a smooth curve that would fit that accurate data. The result was his *Astronomia nova* (*The New Astronomy*), published in 1609, in which he announced his first two laws of planetary motion: (1) planets move around the sun in ellipses (not circles), with the sun at one focus of the ellipse; (2) the area law, that the radius vector (the straight line from the planet to the sun) sweeps out equal areas in equal times. The area law means that when the planet is most distant from the sun in its elliptical orbit (at aphelion) it moves most slowly, whereas when it is closest to the sun (at perihelion) it moves most rapidly. The planet's orbital velocity is not constant, as the ancients had claimed, but varies constantly between the two extremes at aphelion and perihelion. Kepler's third law, the so-called harmonic law, appeared in 1619 in *Harmonice mundi*

(*The Harmony of the World*). It tied the entire solar system together mathematically by relating each planet's average distance from the sun to the time that planet takes to complete one orbit around the sun. If one cubes the average distance for any planet and divides that number by the square of the time that same planet takes to go once around the sun (one year in the case of our earth), then the number obtained will be a constant. The same number is obtained for each and every one of the planets in our solar system. Kepler's three laws demonstrated some quite remarkable regularities in the solar system that no one had suspected before, and the third law especially was essential to Newton's discovery of the inverse-square law of gravity. [4; 5; 10; 19; 40; 52; 73; 102; 106; 107]

Kepler's work was fundamental to Newton's, but it was really Kepler's work as much as anything that demanded a new solution to the problem of planetary motion. As long as it was thought that the planets moved at constant speeds in circular patterns because it was the nature of heavenly bodies to move in circles, astronomers had seldom asked what made them move. But with elliptical motion at variable speeds, one could no longer ignore the problem: in the ellipse the planet moves most rapidly when it is closest to the sun (perihelion) and most slowly at the greatest distance from the sun (aphelion), and its speed varies constantly between those two extremes at all the other points in its orbit. What could possibly make an orbiting body behave like that? Kepler had offered a tentative solution—a magnetic force emanating from the sun—but his solution did not work well at all and was not widely accepted. [5; 10]

It fell to Newton to solve the problem and to put the pieces of the puzzle together in a satisfactory way, synthesizing the terrestrial physics of Galileo with the celestial physics of Kepler to found a universal physics based on the law of universal gravity. Gravity is the attractive centripetal force that pulls the planets toward the sun with an acceleration identical to the acceleration of gravity on earth that acts on falling bodies. The tendency of the planet to fall toward the sun is balanced by its other tendency, to continue at every moment to move in a straight line in the direction it was already moving, the force of inertia. That was, of course, the balance of forces Hooke had suggested to Newton in 1679, but Newton carried the solution far beyond Hooke's speculations by powerful mathematical demonstrations and quantitative laws. [10; 73; 77; 101]

There were three axioms, or laws of motion, upon which Newton based this work: (1) every body will continue in its state of rest or uniform motion in a straight line unless it is compelled to change its state by (external) force impressed upon it (the law of inertia); (2) a change of motion is always proportional to the motive force being applied to the body, and the new motion will be in the straight line in which the force is impressed; (3) for every action there is always an equal and opposite reaction. [77]

From that starting point Newton was able to derive mathematically the laws of his predecessors and go on to develop his own law of universal gravity: the force of gravity is always equal to the constant of acceleration times the product of the masses of the two attracting bodies divided by the square of the distance between them. The earth attracts the moon and keeps it in its orbit around the earth: likewise the moon attracts the earth, an effect most noticeable on the watery surfaces of our globe, where it causes the tides. The sun attracts the planets and their satellites; but likewise the planets attract the sun and the other planets, and so forth, the latter effect being noticeable in certain perturbations (disruptions) in the predicted orbits. Since mass is involved in the general equation, Newton was able, once the forces were known, to evaluate the weight of celestial objects far beyond the reach of human scales. One must suppose even Newton himself was a little awed at what he had done; certainly many of his contemporaries and his later followers were. [77; 101]

The *Principia* offered succeeding generations two primary aspects. It synthesized terrestrial and celestial physics in a way that had never been done before, for as we saw above, Aristotle's system was sharply divided at the sphere of the moon and had one set of physical laws below the moon and a different set above the moon. But the *Principia* also had another side. In addition to its grand unified vision of the universe, it provided a rational mechanics for the operations of machines on earth. The artisans who had created the machinery so common in western Europe by the seventeenth century had had handbooks to guide them, and rules of thumb worked out and passed on in craft guilds, but they had never before had definite physical laws to enable them to calculate and predict mechanical forces and rationalize their control. As we will see in Part II, Newton's followers soon spread the new mechanics to sections of the population that could not read the *Principia* for themselves and thus helped facilitate the Industrial Revolution of the eighteenth century.

But for all the importance of his work, Newton no longer had a mechanical explanation for gravity, and that was a serious problem. Continental natural philosophers accused him of reintroducing occult qualities into natural philosophy, meaning that attractive centripetal (center-seeking) force of gravity, because there was no physical, material substrate to explain the action of gravity. Such occult (hidden, secret, nonphysical) qualities and forces were exactly what the seventeenth-century mechanical philosophers had hoped to eliminate with their aethereal explanations, where matter operated on other bits of matter by impact or pressure. Human beings can understand those sorts of contact mechanisms, for it is ever so easy to envision them; we see such contact forces around us every day. But it is impossible to imagine an attractive force of gravity, without physical contact, reaching across millions of miles of empty space to hold a planet in its orbit. So in a sense the Continental philosophers were quite correct in saying that Newton had reintroduced occult qualities into natural philosophy; Newton really was no longer an orthodox mechanical philosopher in the seventeenth-century meaning of the term. Newton had his mathematical laws to explain the action of gravity, and that was all he had. He had no physical explanation of it at all, and that left him very uneasy, for he had previously accepted the argument that all force must be exchanged by some sort of contact mechanism. [18; 101]

After the *Principia*

The writing of the *Principia*, which was to become the foundation of modern science, did not make Newton himself into a modern scientist. His concerns remained centered primarily on religious issues, though he was, after the great success of the *Principia*, more willing to think that True natural philosophy could contribute to the goal of rediscovering the True religion once known to the earliest peoples.

Newton began to compose a treatise entitled *Theologiae gentilis origines philosophicae* (*The Philosophical Origins of Gentile Theology*) in the early 1680s and continued to work on it for a number of years, though it remained unpublished. In it he said that the ancients had known the true religion before they corrupted themselves and became idolatrous (worshipping idols and false gods). Working his way backward through the several idolatrous traditions that had flourished among the Egyptians, the Chaldeans, and the Assyrians, Newton found behind the idolatry the uncorrupted religious Truth that had once been universally recognized among them. In addition, attributing to the ancients an understanding of Copernican heliocentrism, Newton said they had also known the true system of natural philosophy in which "the fixed stars stood immovable in the highest parts of the world; that under them the planets revolved about the sun," and that the earth, as one of the planets, "described an annual course about the sun." That the ancients had known uncorrupted Truth in both religion and natural philosophy was fully consonant with Newton's general stance on primitive wisdom and offers no surprise. But in the often reworked *Origines* manuscripts Newton carried his convictions somewhat further. [18:750; 100; 101]

With their understanding of true natural philosophy, the ancients were enabled to create a form of religious structure and worship that adequately represented to human beings the structure of God's cosmos and suggested the study of nature as a means to satisfy human aspirations for knowledge of God—an extremely rational way of going about things, Newton thought.

So then the first religion was the most rational of all others till the nations corrupted it. For there is no way [(wth out revelation), *interlineated by Newton in the original*] to come to ye knowledge of a Deity but by the frame of nature. [18:151]

Newton spoke there as an early modern natural philosopher imbued still with Christian doctrines of long standing: nature being theologically transparent, a knowledge of "the frame of nature" led directly to a knowledge of the Deity and was indeed the preferred route to follow in the post-Reformation turmoil over valid interpretations of revelation. The universe having been created by the Deity according to Judeo-Christian tradition, the created world reflects the attributes of its Creator, just as a work of human art reflects the character of the artist. The Psalmist had said, "The heavens declare the glory of God" (Psalms 19:1), and St. Paul had added the classic Christian version: "For the invisible things of him [God] from the creation of the world are clearly seen, being understood by the things that are made, *even* his eternal power and Godhead" (Romans 1:20). [18; 42]

It is indeed the case that a few (but very few) ancient philosophers and astronomers had suggested that the sun rather than the earth was the center of the world. Newton treated them as being closer to the uncorrupted Truth in natural philosophy, though that is not how modern scholars understand the beliefs of antiquity, for most ancient thinkers of which we have any record at all believed the earth to be central. Newton, as a late Renaissance thinker, however, continued to believe that the most ancient had been closest to original pure and uncorrupted knowledge. [18]

But a great many thinkers from antiquity did certainly attribute special value to the sun, perceiving it to be the source of warmth, light, and life, and often identifying the sun as a god. Many religious ceremonies did also take place in a sacred space in which a sacred fire burned in the center; perhaps the sacred space did represent the world and the sacred fire did symbolize the sun and its associated deity for those peoples. In any case, that was what Newton thought he understood of the most ancient religions he had studied. The structure by which the ancients represented the world in the most ancient form of religion, Newton said, was "a fire for offering sacrifices [that] burned perpetually in the middle of a sacred place." This arrangement, which Newton called a prytaneum, symbolized the cosmos, with the fire representing the sun at the center and the sanctified space around the central fire representing the entire world, which was "ye true & real temple of God."

> The whole heavens they recconed to be ye true & real temple of God & therefore that a Prytaneum might deserve the name of his Temple they framed it so as in the fittest manner to represent the whole systeme of the heavens. [100:24]

Thus, after the success of the *Principia* in establishing (or reestablishing) a heliocentric astronomy and natural philosophy, Newton thought he had taken a firm step toward his goal of reestablishing the true religion of the ancients. The ancient religion was the most rational of all, he had said, and it encouraged humanity to gain knowledge of the Deity "by the frame of nature." The true "frame of nature" now being set out in the *Principia,* human beings could at last return to the true religion that was based on the true natural philosophy of heliocentrism. For Newton all his study of nature had that religious motivation. [18; 100; 101]

Newton was to live for another forty years after 1687, and he seems to have taken seriously the program of restoring true natural philosophy, which he had himself set forth in the *Origines* papers, as the best way of restoring true religion. In his later work in natural philosophy three principal areas of concentration may be detected. First and foremost came Newton's continued pursuit of alchemy. Next came an intense concern to find the cause of gravity. Finally, Newton made an immense effort properly to incorporate comets into his new system of the world. One should not assume, however, that Newton's interest in theological issues had in any way declined even though he shifted the focus of his study (for the most part) away from the written records of Scripture, church fathers, sacred ritual, church history, and so forth. He had, after all, explored all of that exhaustively already. Perhaps because of the remarkable success of the *Principia* itself in restoring true natural philosophy, Newton shifted his focus to *more* study of natural philosophy as the *best* way to restore true religion, but nevertheless theological and religious issues present themselves vigorously in every remaining major episode. [18]

Newton returned to his alchemical experimentation as soon as Book I of the *Principia* was dispatched to London, in April 1686, and alchemy absorbed much of his time and energy. Laboratory notes written between that date and the date of his departure for London and the Royal Mint in 1696 comprise close to 55,000 words; his other alchemical papers of the same period, an estimated 175,000 words. He shared many of these interests with a new friend, Nicolas Fatio de Duillier (1664–1753), a young Swiss mathematician who flashed like a brilliant meteor into Newton's life shortly after the *Principia* was published. Newton and Fatio developed an unusual intimacy and even spoke of living in close proximity. Their shared work seems to have focused on the interpretation of ancient alchemical literature and its hidden wisdom and also on experimental signs of life in the mineral kingdom, but they did in addition some work on the interpretation of biblical prophecy. [18; 66; 71; 101]

These foci were closely related aspects of Newton's search for evidence of God's action in the world, but it is not so clear that Fatio ever fully understood or participated in the alchemical part of the search. Most of the evidence for their joint alchemical activities comes of necessity from Newton's papers, for Fatio did not leave much in this area; their relative degree of involvement in alchemy is probably accurately reflected by that disproportion. In the interpretation of ancient biblical prophecy, on the other hand, the two men were probably closer together for a time: their mutual interest in both alchemy and prophecy was most intense in the late 1680s and the 1690s. But Fatio's involvement with prophecy became ever more consuming. Leaving the dry bones of academic interpretation behind, Fatio heard the voice of God directly from the French Prophets (refugee French Huguenots) who arrived in London in 1706. Newton, reportedly restrained by other friends from attending the public prophetic sessions, discussed the meaning of the new prophecies with Fatio as late as January 1706–07. After Fatio became in effect a recording secretary for the inspired group, however, he probably no longer had time for alchemy, whereas Newton continued to labor over his alchemical papers occasionally and to incorporate alchemical ideas into other aspects of his work. [18; 48; 93; 101]

With respect to a causal principle for gravity, Newton found himself in a difficult position in the *Principia* and in the post-*Principia* years. Having believed for some 20 years in a material aether as a mechanistic cause for gravity, he was forced to abandon that belief because of its incompatibility with actual celestial motions. A combination of mathematics, observation, and experimentation had forced him to give up the mechanical causal principle he had accepted for so long. What was he to put in its place? Two tentative solutions to the problem of a cause for gravity emerged in Newton's later years. One was that the omnipresent Supreme Deity subsumed gravity directly. The other was that an intermediate agent existed between God and the world, an agent that could account for gravity (as the vegetable spirit, an intermediary between God and the world, accounted for vegetation), but the agent for gravity had to be of such a nature that it would not constitute a drag on the motions of the heavenly bodies.

As noted above, Newton went on record as an undergraduate with his belief that God was everywhere present—where space was empty of body but also where body was also present. He never seems to have wavered in that conviction, which was a fairly conventional Judeo-Christian one.

But the business of linking God's literal omnipresence to gravity involved Newton in much study of ancient sacred and secular literature. As we have already seen, Newton had always accepted the Renaissance view of history as a decline or a falling away from an original golden age, a time in which there had existed an original pure wisdom of things both natural and supernatural, an ancient wisdom subsequently lost or garbled through human sin and error and through temporal decay. By the 1680s Newton had already been engaged for a long time in attempts to restore the original truths once known to humanity by decoding obscure alchemical texts and by searching ancient records for the original pure religion. So when the (for him, modern) mechanical explanation of gravity failed, it was only natural that he should turn to ancient sources in an attempt to recapture the truer explanation of gravity once known to the wise ancients. What he found there in antiquity he then incorporated into his cosmic system. [18; 21; 92]

Newton reviewed the sacred writers, later citing passages from the most ancient Hebrew authorities through earliest Christian antiquity: Deuteronomy, Kings, Psalms, Job, Jeremiah, John, Acts. No doubt the sacred writers reinforced his sense of God's presence everywhere and of God's great power, but they were silent on Newton's profound question of *how* God's omnipresence might directly subsume the mathematical laws of universal gravity he had found. [18; 42; 77; 92]

On the other hand, the ancient natural philosophers were not entirely silent on the question of God and gravity. Newton told a friend in 1705 that they did believe God was the cause of it. We do not have a precise record of all he read in search of an answer to his question, but manuscript and published materials do indicate some of his sources, and they ranged from pre-Socratic Greek philosophers of the sixth century B.C. to Aratus, a Greek Stoic of the third century B.C. who wrote on astronomy, through Cicero and Virgil of Roman antiquity, to

Philo, a Hellenistic Jewish writer who died in A.D. 50. Newton also studied with some care the musical myths of the ancients, such as those about the god Pan (master of the pan-pipes) and Orpheus (master of the lyre). And, as with his work on the origins of gentile theology, Newton's search for ancient opinions on gravity led him to make eclectic, selective use of authorities and to mold them into a coherent whole that spoke to his own purposes. [18; 21; 32; 34; 41; 46; 54; 59; 61; 86; 87; 88; 89; 90; 92; 96]

The coherent whole that Newton created from his readings was a sort of spiritualized version of Stoic philosophy in which the divine Stoic *pneuma* (or *spiritus* or aether), which had originally been material in nature, was completely dematerialized to become totally incorporeal but was still a divine substance that permeated and mingled with everything in the cosmos and that constituted the ground of all being for the cosmos. The Stoic *pneuma* had become a Divine Mind, immediately present to every particle of matter in the universe, guiding the operations of matter from moment to moment, and always intelligently aware of everything in the universe. The later Stoics that Newton read during this period of his life did sometimes dematerialize the Stoic *pneuma* in that fashion, and Newton found the concept easy to reconcile with the Judeo-Christian tradition of God's omnipresence and His omniscience (having knowledge of everything), as well as with the myths about the gods Pan and Orpheus, deities who had controlled matter through musical harmonies that were based on mathematical law, just as his own law of gravity was. [18]

From the Stoics during this period he also took their explicit dichotomy of "passive" and "active": the passive principle was always matter, whereas the active principle was always that which acted upon matter and was variously called by the Stoics "Cause," "Mind," "Reason," or "Force." The active principle was always in some sense divine, and the dichotomy of passive and active that Newton then adopted served him in the same way his earlier dichotomy of mechanical and vegetable had done. But whereas gravity had definitely been mechanical in the 1670s, when he first formulated that earlier dichotomy, it had now moved to the other side of the aisle to become active and divine.

In 1706, in what was then Query 23 of the *Optice* (the second edition of the *Opticks* of 1704, to which new material had been added and the whole translated into Latin) and in what was to become Query 31 of the next edition of the *Opticks*, that of 1717–18, Newton openly designated gravity as an active principle in his most general statement on the forces associated with matter, the particles of which he described as "solid, massy, hard, impenetrable, [and] moveable."

It seems to me farther, that these Particles have not only a *Vis inertiae* [force of inertia], accompanied with such passive Laws of Motion as naturally result from that Force, but also that they are moved by certain active Principles, such as is that of Gravity, and that which causes Fermentation, and the Cohesion of Bodies. [75:400–401]

Within the context of Stoic doctrines, the divinity of Newton's active principles is perfectly obvious. Defined against the passive principle of matter, they are the divine eternal Cause, Mind, Reason, Force. [18]

Newton's life of retiring scholarship had ended in 1696 with his appointment as Warden of the Mint just as the great recoinage of William III's reign got underway. At that time the principal duties of the Warden were of a legal nature, and Newton was expected to oversee the detection and prosecution of counterfeiters and clippers of coins. He superintended that work in detail and also dealt from time to time with other infractions, such as conflicts between Mint personnel and the garrison of the Tower of London. In 1699 he was named to the Mastership of the Mint, which position he held until his death. The Mastership was the general administrative post of the organization, and Newton handled it with his accustomed thoroughness, straightening out the chaos of records and accounts left by his predecessors and, when questions of procedure arose, searching out all the precedents as far back as Elizabethan times. The Mastership of the Mint cushioned Newton's later years with a financial security he had never known before, and he died a wealthy man. [17; 101]

Honors accumulated for the aging Newton: in 1699 he was elected as one of the first eight foreign associates of the French Royal Academy of Sciences; in 1703 he became president of the Royal Society of London, to which office he was reelected annually until his death; in 1705 he was knighted by Queen Anne. [17; 101]

It was as he sat in the Presidential Chair at the Royal Society that Newton began to catch a glimpse of his second solution to the problem of a cause for gravity. Newton had spent an immense amount of effort ransacking the complex ideas of the ancients on the matter of gravity, and he seemed for a time to have been convinced that the Deity was directly responsible for the operation of gravity, with the divine immaterial substance permeating all the space between bodies and indeed permeating bodies themselves. That solution to the problem of a cause for gravity constituted a resounding answer to those atheistical tendencies inherent in the mechanical philosophy detected so many years before by Henry More, Isaac Barrow, and Ralph Cudworth, and under their tutelage by Newton himself, for it caused God to be most intimately involved with all events in the natural world. [18; 39]

However, for an Arian theologian, which Newton certainly was, that solution would have generated considerable unease. According to Arian doctrines the Supreme God should not be responsible for the moment-to-moment movements of all the particles of matter in the universe, because the Supreme Deity in Arianism was much removed from the created world, so high above it (transcendent) that He could not interact with it directly.

Arius had rejected the argument that the Son of God, Jesus Christ, was of the same "essence" as God the Father, and in so doing Arius had reemphasized the absolute supremacy of the original uncreated One Who had existed from all eternity. Arius insisted that there was a period before the creation when the Son had not yet been called into existence. The status of the Arian Son was necessarily diminished as the status of the Arian Father was elevated: Father, Son, and Holy Ghost did *not* share as coequals in the Arian Godhead. The Son was the first and most important of created beings, even though he was not granted full equality with the Father, and in Arian doctrine he became the mediator or intermediary between God and the created world. One result of Arius's emphasis on the supremacy of the Father was to set Him wholly apart from the creation, to make Him entirely "other" and transcendent. Consequently, the cosmological role of the Arian Christ was also given emphasis. The Christ was the intermediary through whom and by whom God created the world and interacted with it. That was what Newton had come to believe in the 1670s, and Newton said repeatedly, in the years after that, that the Supreme God does nothing by Himself that He can do by others. Arian theology was undoubtedly the reason, then, that Newton was uneasy about his first solution to the problem of a cause for gravity. Wanting his partial Truths from all fields of study to coalesce, he attempted to construct a new solution to the problem of a cause for gravity. [18; 28; 101]

The difficulty that Newton faced, the same difficulty he had had since 1684, was the issue of corporeality or materiality. No aether made up of the corpuscles or particles of ordinary matter would do for his theory regarding an agent causing gravity, because such an aether would create friction as celestial bodies made their way through it, and that would slow them down. But celestial bodies showed no sign of such frictional retardation in their motions. Could there, on the other hand, be some other kind of substance that would constitute a nonretarding aether? A substance intermediary between the full corporeality of ordinary matter and the full incorporeality of the spiritual divine substance of the Supreme Deity? A sort of "spiritual body"? Newton's studies in both biblical and alchemical texts had introduced him to that term "spiritual body," but where did such a thing exist in the world of everyday experience? The answer came to Newton as he sat in the Presidential Chair at the Royal Society and watched a long series of experiments and demonstrations on electricity by Francis Hauksbee (1670–1713), who was then a Fellow of experimental demonstrator at the Royal Society. [18; 38; 39]

Electricity had certain qualities that seemed to belong to bodies. For example, it was tangible, that is, it could be perceived by the sense of touch. Newton reported that he himself had felt it. On the other hand, it did not seem to weigh anything, and Newton had definitely demonstrated in the *Principia* that all bodies weighed something in relation to other bodies; their mass was a part of the equation of universal gravitation. Yet, Newton noted, the electric "effluvium" or "spirit" could be agitated and emitted from a body so as to show its effects two feet away from the body, and the body lost no weight. So here, if anywhere in the world, was a substance that seemed to Newton to partake of the qualities of both the spiritual and the physical.

This growing recognition of the properties of electricity between 1703 and 1708 stimulated Newton's fertile brain enormously. Electricity was associated with light in Hauksbee's work, and Newton, who had long associated the corpuscles of light with the vegetable spirit of alchemy, at first thought that electricity might be the answer to the long, long search he had made for the agent God used to organize the small particles of bodies. His private papers show that he speculated on that possibility, but he soon saw that the properties of electricity might be the answer to his difficulty about gravity. [18; 29; 35; 36; 37; 43; 44]

Fixing his attention on the property of elasticity that electricity seemed to have, that is, its ability to expand by repelling itself away from its source and by its particles mutually repelling each other, Newton constructed a new sort of aether to account for gravity. Newton argued that this new aether was so elastic that it filled the heavens by the efforts its particles made to recede from one another; he also argued that it offered no resistance to the planets and comets moving through it. The new aether was active and not mechanical in its operation, so it seems most unlikely that Newton intended it to be a concession to the orthodox mechanists who had criticized his lack of a mechanism for gravity. Quite the contrary, Newton seems to have constructed his new active aether in response to his own theological anxieties regarding the immense involvement with matter that he had earlier postulated for the Supreme Deity. Since Newton thought that the Supreme God does nothing by Himself that He can do by others, an intermediary agent for gravity would have seemed more appropriate to him. By definition, it was an agent closely assimilated to divinity, by virtue of its inherent activity, but to an Arian theologian it was more acceptable simply because it was an *agent*, a created being that acted for the Supreme God in the moment-to-moment supervision of matter. [18; 36; 44; 63; 75; 101]

Regarding comets, Newton had long evinced substantial interest. Traditionally, comets had always seemed to most people to be wild danger signs, signals that some sort of disaster was about to strike. Comets appeared and disappeared in an unpredictable way and often looked like fiery swords cutting across the normal circular patterns of the heavens. Newton recognized, however, in the early stages of writing the *Principia*, that comets are subject to the inverse-square law of attraction toward a center of gravitation. By the time the *Principia* was published in 1687 Newton knew that comets describe conic sections (regular mathematical figures, explained below)—conic sections focused on the sun. The taming of the comets, making them more or less domesticated members of the solar system, was not the least of Newton's achievements in the *Principia*. [18; 101]

The conic sections are cuts or "sections" made of a cone. They are four in number: (1) the circle, a cut parallel to the base of the cone; (2) the ellipse, a cut of the cone made at a shallow angle; (3) the parabola, a cut parallel to one side of the cone; (4) the hyperbola, a cut made at a still steeper angle. Of these, only the circle and the ellipse make closed curves that enclose a definite area. The other two are not closed curves. So if a heavenly body, such as a comet, described a parabola or hyperbola as it moved around the sun, it would not repeat that orbit in the future but would fly away into outer space. If, however, the conic section described by a comet is a circle or an ellipse, the orbit will be closed and the comet will make a periodic return.

Newton and Halley both worked strenuously on the orbits of comets, especially after the spring of 1694. Newton had a graphical method for computing parabolic approximations to closed elliptical orbits, but it was important to both men to demonstrate definitively that the highly eccentric (off-center and therefore noncircular) orbits of their comets were truly elliptical and truly closed. Only then could comets become permanent members of the solar system. "Halley's Comet," still famous, was that of 1682; its period was established at about 75–76 years. Its return has now been duly recorded four times and since 1758 has often been cited for the definitive establishment of Newtonian principles of celestial dynamics. We will learn more about the significance of Halley's Comet in Part 2. [18; 71; 73; 77; 91; 99; 101]

Newton's favorite comet, however, was the comet of 1680, which he

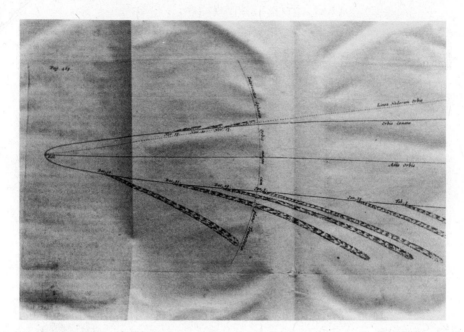

FIGURE 1.7   A graphical representation of the path of a comet
published in Newton's *Principia*. Courtesy of the University of Pennsylvania Library, Rare Book Collection.

and Halley concluded had a period larger than 500 years. It has not
yet returned, but Newton thought he had found a special function for
that particular comet. Newton fully realized that, according to his own
concept of universal gravitation in which all bodies *mutually* attract
each other, comets might very well be sufficiently disrupted over a
long period of time by passing by the several planets that they would
fall into the sun rather than pass around it. He thought that was probably the ultimate destiny of the comet of 1680. It might have five or
six more revolutions first, but, if and when it did fall into the sun, it
would increase the heat of the sun enough to burn up the earth. Always, as usual, attempting to get his partial Truths to coalesce, Newton
thus thought the comet of 1680 would be the natural agency God
would use to usher in the destruction of this world predicted in biblical prophecy. [13; 18; 24; 25; 91; 101]

The Impact of Newton's Work

Newton's system of the world, based on mathematical law and order but under the supervision of a providential Deity, was readily adopted by moderate churchmen and laity in the Church of England. Especially after the Glorious Revolution of 1688–89, when the Catholic James II was deposed and a new balance achieved in the English constitution, the English church found the new balance in cosmic forces and motions to serve its social and political needs admirably, as we will see in Part 2.

The Boyle Lectures, instituted by the last will and testament of Robert Boyle, who died in 1691, then provided a forum for spreading Newtonian natural philosophy to a broad segment of educated London. Early Boyle lecturers included Richard Bentley (1662–1742), a classical scholar who later became Master of Trinity College, Cambridge; Samuel Clarke (1675–1729), translator of Newton's *Opticks* into Latin, spokesperson for Newton in the Leibniz-Clarke correspondence (discussed below), and, like Newton, an Arian in theology; William Whiston (1667–1752), another Arian, Newton's successor in the Lucasian Chair of Mathematics, and later a lecturer on Newtonian mechanics; and William Derham (1657–1735), who later published two very popular works, *Physico-theology* (1713) and *Astro-theology* (1715), that were based on his Boyle Lectures of 1711–12. These disciples of Newton were dedicated to showing how the Newtonian system demonstrated the wisdom and providence of God, justified order and stability in the polity, and supported the contemporary balance in the constitutional settlement. Newton was well pleased. He wrote to Bentley late in 1692, "When I wrote my treatise about our Systeme I had an eye upon such Principles as might work with considering men for the belief of a Deity & nothing can rejoyce me more than to find it useful for that purpose." [71: v. 3, 233] [See Part 2]

But faith was not ultimately the use to which Newton's system was put. In a sense the world had passed Newton by in his own lifetime, and the concerns with religion and with finding a unified system that encompassed both natural and divine principles, so very strong in Newton himself, were of much less concern in the eighteenth century. The religious wars and political upheavals of the sixteenth and earlier seventeenth centuries that had constituted so much of Newton's own heritage gave way in the later seventeenth and eighteenth centuries to a period of relative stability in European social and political life—at least until the great democratic revolutions began in the later eighteenth century—and during that period of relative calm the interest and focus of human activities shifted from religion to practical and utilitarian concerns. The people of the European nation-states struggled to prosper in the here and now, not worrying quite so much about the correct route to a blessed hereafter. The newer attitudes promoted the practical and utilitarian applications of the principles of Newtonian mechanics to earthly machines. [See Part 2]

One may look very briefly at two other significant developments that served to undermine the religious side of Newton's system. The first of these began with the famous correspondence between Samuel Clarke and Gottfried Wilhelm Leibniz (1646–1716), codiscoverer of the calculus and a German philosopher and diplomat of considerable international standing. A bitter priority dispute over the discovery or invention of the calculus had erupted earlier between Leibniz and the Newtonians, who claimed that Leibniz had learned his method from papers Newton circulated in manuscript, so that even though Leibniz published first, Newton should get full credit for prior discovery. Modern scholars have concluded that Leibniz and Newton made their discoveries independently and both deserve full credit, but the bitterness of the priority dispute was very much still present as background to the Leibniz-Clarke correspondence of 1715–16. The two men exchanged five letters each through writing to Caroline, Princess of Wales, a former pupil of Leibniz, and their letters were soon published and widely read. [1; 33; 101]

Leibniz raised a number of significant objections to the Newtonian system in his letters, but the point of greatest interest in the correspondence, in view of what happened to Newton's system in the eighteenth century, had to do with what sort of world-machine the Deity had created. Newton certainly had focused strongly on the regular operations of the laws of gravity in running the world-machine, but because of the mutual influence of many bodies on each other he had not supposed it would run forever in its present order. He had thought that when the perturbations in the solar system, for example, grew sufficiently great, the system would need a reformation, and at that time God would step in to set it in order again. Or, if God's plan called not for reformation but for the destruction of the world, that could be effected through the regular operation of the laws of gravity on the comet of 1680. The important thing for Newton was that the world was constantly under divine supervision and would proceed as God saw fit. Leibniz sarcastically observed that Newton's God seemed not to have had enough foresight to make the world-machine run right perpetually, that Newton's God was an unskillful Workman Who had to keep mending His work. Clarke, who is generally understood to have been Newton's spokesperson in this debate, argued that treating the world as a great machine that would go on running without God's intervention was aligned with materialism because it tended to push God's providential government out of the world. [1; 18; 33; 101]

The ideas Leibniz expressed on this point were hateful to Newton; he had fought against them all his adult life. But the very successes of Newtonian principles soon showed that the world was not running down: the perturbations in the solar system cancel out so that the system is stable and God's intervention is not required. Indeed it was the *regularity* of the Newtonian system that most captivated eighteenth-century thinkers, especially after the regular operation of gravity brought the return of Halley's Comet in 1758 and dramatically confirmed Newton's mathematical principles. The eighteenth century focused more and more on the regular mechanical operation of natural laws that did not require divine supervision. So in the end Leibniz won what he feared: Newton's active and providential God soon became the remote and distant God of the deists, a Deity Who might have created the world originally but Who then left it to run by itself. [18; see also Part 2]

Finally, there is the fate of Newton's theory of matter. Newton had insisted that matter itself was passive and that the forces in nature— the cosmic force of gravity, and the forces of repulsion and fermentation and cohesion in micromatter—were active principles (by which he meant divine, as we have seen). The active principles operated only between and among the particles of passive matter and were *not* attached directly to matter itself. To incorporate the forces into matter would make matter active in its own right and would make it independent of the Deity. Such an active matter would be self-starting and self-organizing and would have no need of divine powers to set it in motion or to guide its small particles into complex forms. A concept of an active matter would lead straight to atheism, in Newton's opinion. Eighteenth-century philosophers challenged Newton's matter theory in the most fundamental way, pointing out that the human sensory apparatus can really tell us nothing about the hard, massy, impenetrable, passive particles of matter postulated by Newton. Our senses cannot even tell us that they exist. All we can really know is that the force of repulsion makes matter seem hard and impenetrable to us, the force of cohesion resists our efforts to take material objects apart, and so forth. For many eighteenth-century thinkers, then, the very concept of matter came to be centered on the forces that Newton had tried to keep separate from it. The result was just as Newton had predicted: the concept of an active matter with forces incorporated into it did lead straight to atheism, the most notorious case being that of Julian de La Mettrie, whose *Man a Machine* in 1748 shocked even the French *philosophes*. [18; 60; see also Part 2]

In conclusion one may say that the work of Isaac Newton served many different functions: for himself, for his own society, for the science of the centuries to come. His own private goals fell away as the world changed, and although his system supported religion in some times and places, his mathematics, his optics, his mathematical principles of natural philosophy, and his rational mechanics, soon largely shorn of their religious and theological substructure, came to stand alone and proved nevertheless to be of inestimable value for subsequent generations.

The Culture of Newtonianism (1687-1800)

Background: Master and Followers

THROUGHOUT HIS LONG life Newton fixed his extraordinary genius upon scientific problems that encompassed the heavens and the earth. As we saw in Part 1, his deeply religious vision was tempered and aided by intense experimental activity, yet it was always and unrelentingly cosmic. His mind roamed through the terrestrial movement of bodies, some as simple as pendula, and then went on to the trajectories of the planets. Given his extraordinary range of interests, the gaps between Newton, the master, and his contemporary followers, seem daunting. The inner workings of his mind, in contrast to the everyday, ordinary world of the other mortals around him, were different—not completely unknowable, but distinctive and even unique. With that difference in mind we can ask, Was Newton a Newtonian, and did his values permeate the new culture of Newtonianism that emerged in his lifetime? In the next breath, we can answer, Of course not. What interested many of his scientific contemporaries and followers passed him by. Scientific genius does not often dwell upon the political situation of a church or upon the stability of the state. But even genius cannot remain remote from its setting.

Newton's setting, which included civil war and two revolutions, perhaps required both distance and engagement. No one among Newton's contemporary followers probably ever knew the range and extent of his intense theological, religious, and alchemical interests. None among them possessed his genius or probably his capacity for work and study. In his heart Newton was a loner. Even though he stood for Parliament as a supporter of the Revolution of 1688–89, few knew the depth of his passionate hatred of monarchical absolutism and of the attempt by James II to bring Catholics into high positions at Cambridge. The depth

of Newton's hatred would have to wait until the discoveries made by
modern Anglo-American research. Only in the 1970s and 1980s, by
delving into Newton's manuscript writings, have modern historians come
to see the intensity of Newton's passions and interests and to relate
them to his political and cultural setting. Again, as we saw in Part 1,
the Newton behind the *Principia* (1687)—metaphysician, antitrinitarian,
even alchemist—has been belatedly revealed largely through his private notations and treatises.

The differences and similarities between Newton and the Newtonians
point us toward the profound changes that had occurred in England
during Newton's lifetime, from the 1640s to the 1720s. It is significant
that in the 1690s all the Newtonians who latched onto Newton (after
he became famous) were men of the younger generation, born after
1660. By contrast, Newton was born in 1642, in the eye of the greatest
political storm in England's history. In that year relations between the
king and Parliament thoroughly deteriorated, and both sides raised
an army to launch what became a civil war.

Into the political crisis of the 1640s religion was woven like a pattern in a tapestry. The rank-and-file supporters of Parliament were rigorous Puritans, devout Protestants schooled in Calvinism. They wanted
to purge the English church and court of the last vestiges of Roman
Catholicism. King Charles I was the kind of Anglican who loved the
splendor and ceremonies of the old pre-Reformation religiosity. He
also craved the power that the Catholic kings in France and Spain,
unburdened by parliaments or fractious Protestants, exercised. Perhaps
civil war between king and Parliament was inevitable. In 1649, when
Newton was a mere boy of seven, the leaders of Parliament put King
Charles I on trial, convicted him of treason, and executed him.

After regicide came the republic of the Puritans. But these rigorous
and godly revolutionaries were challenged on every side. Within this
first English revolution also unfolded another, more radical and plebian
movement that sought to redress the deep social and economic divisions within English society. Newton was an adolescent in the 1650s, a
period of crisis: war abroad against the Dutch; purges of Parliament;
fear of lower-class revolt led by radical Protestant sects; political shift-
ing from left to right as the landed classes tried to figure out how a
country that had known only monarchy could be governed without
one. With the death of the one man who seemed able to hold it all
together, Oliver Cromwell, the country drifted, until finally the army
moved to restore the king in the person of Charles II, the exiled son
of the executed king. In 1660 the restoration of monarchy occurred
with widespread approval; Newton was just turning 18 years old. [41]

In the depth of his youthful passion to know God, Newton resembles the Puritans. In his mature writings on ecclesiastical polity where he advocated as broad and tolerant a church as possible, he resembles someone who has seen religious civil war at close hand and wants never to see it again. In his fear and hatred of any philosophy that leads to atheism, Newton also resembles all the churchmen and Puritan ministers of his age who attacked the radical Protestant sectaries of the Revolution, who accused them of being atheists because they believed that the distinction between God and his creation should be rethought, if not banished. The radicals believed that the chasm between God and man justified the abusive power of priests and clergymen. Because of the gap between Creator and creation they alone claimed to be mediators between the divine and the human. To abolish the gulf and to undermine the clergy, the radicals revived ancient heresies and said that God was one with the creation. Radicals like the Digger Gerrard Winstanley and the Leveller Richard Overton preached a kind of pantheism. Many of them also believed the world to be near its end; then, at last, in the millennial paradise the meek and the poor shall inherit the earth. This God-in-Nature pantheism amounted to atheism in the mind of Newton and the clergymen who taught him in Cambridge. We now might call it naturalism or materialism.

But despite his fear of materialism, in other ways Newton was like the radicals. In believing that the world would someday end, they and Newton also resemble every devout Protestant of the mid-seventeenth century, from Archbishop William Laud to John Milton, Oliver Cromwell, and the radical Quakers and Diggers. Whether sectarian radicals or political and religious moderates, they all believed that history and time are finite things, that the world began at an appointed moment and so too will end when God wills the apocalypse to happen. All these beliefs Newton could have heard expressed from the pulpits and meeting rooms of the 1640s and 1650s. The revolution of midcentury was framed by a distinct understanding of time: a biblical time with a definite moment of beginning, revealed in the Book of Genesis, and then a promised end, foretold in the Book of Revelation and other biblical passages. [35] Time itself came from the divinity; its realness was signaled by its having a literal beginning and ending. In Newton's vision time was as an actual arena where moving bodies coursed their trajectories. Newton wove that notion of absolute time into his physics; it became part of its metaphysical foundations.

But by the early eighteenth century, even to many of Newton's closest followers, real, biblical time was beginning to seem less credible. The English began to think differently about this world, and they did so for reasons that were both economic and political. Very gradually, as prosperity increased, time began to seem unbounded, an infinity without limits, one relative to circumstances, events, interventions, a shapable time. Although Newton lived on well into the eighteenth century, his mind and his metaphysics, as we saw in Part 1, belonged to the seventeenth. Gradually his religious and metaphysical vision slipped out of the thought, values, mechanical practices, even the religiosity, that we label the culture of Newtonianism.

By the 1690s the world of instability and relative paucity that Newton knew as a child and young man had slowly begun to disappear. By the end of the century a new agricultural prosperity affected large parts of southern England. So rich and plentiful were the harvests that the English, in striking contrast to the French, became net exporters of grain. In much of England famine and subsistence crises became phenomena of the past. [1] In precisely this period, decades of political instability brought about by conflict between crown and Parliament ended. In 1688–89 a final revolution eliminated both absolutism and Catholicism as viable forms of political organization or worship. By the 1730s prosperity translated into consumption, and the domestic market for goods and services vastly expanded. [52]

In 1689 Parliament expelled King James II, a Catholic and son of the long-ago-beheaded king; they managed to effect this revolution without bloodshed. In his place Parliament appointed William of Orange, the *stadholder* of the Dutch Republic. His wife was the daughter of James II, and with her Protestant husband leading a Dutch army, supported by English coconspirators, she watched her father's exile without shedding a tear. Official Catholicism and absolute monarchy, a system of government still found on the Continent, disappeared in England. In its place English theorists described their king-in-parliament government as having been providentially ordained. God approved of a constitutionally bound monarch who ruled with Parliament, both inspired by providential guidance and approval. In reality eighteenth-century Britain was ruled by an oligarchy of families with deep roots in the land but with side roots out into commerce and industry. Not only were their representatives sitting in Parliament, but Providence watched over them from the Newtonian heavens and all was right—at least for them in their world.

Thus within Newton's lifetime not only did famine largely disappear; so too did absolute monarchy, replaced by an oligarchic king-in-parliament form of government. With it came religious toleration for all Protestants and relative freedom of the press. Government was now to be financed by loans as well as taxes, thus giving the landed and propertied a stake and a say in policy, a degree of power that only the Dutch elite in their republic could rival. After 1695 the censorship of books loosened to the extent that a publisher or author could only be prosecuted *after* a book had been published. The office of censor, vetting and excising manuscripts submitted by would-be authors, now became an institution found only in absolutist countries. Foreign travelers marveled at English prosperity, stability, mechanical skill, and relative religious and political freedom; by the 1720s they also observed that in England scientific culture had come to be entirely dominated by the Newtonian synthesis. Even English Protestantism had been deeply affected by Newton's achievement.

The First Newtonians

In the 1690s the English educated elites possessed a new and extraordinary cultural resource: Newton's own science. No person ever controls the thoughts that he or she puts into the world through books. The first generation of Newton's followers, almost all of whom were clergymen, tried to use Newton's legacy as he wanted. But it did not always turn out that way. Newton wanted his natural philosophy to show the power of the Creator, and indeed for some of his clerical followers like Richard Bentley and Samuel Clarke, it did. But Newtonian science also had very practical, mechanical applications easily extracted from it. Among Newton's more utilitarian followers such as Jean T. Desaguliers (b. 1683), his mechanics of terrestrial bodies in local motion became applied, both technically and conceptually, to problems in mining, drainage, and pumping. [67] Bentley and Desaguliers may be taken as typical of the first generation of Newton's followers, the Newtonians. They were neither original thinkers nor original scientists, yet as preachers and mechanists they, and the Newtonians who came after them, shaped more than a half century of British religious thought and practical mechanical applications.

Both Bentley and Desaguliers learned Newtonian science partially through direct contact with the master. Although Newton withheld from public scrutiny his beliefs on a variety of subjects, he should not be portrayed as a recluse. The *Principia* had made him famous. Long before his death in 1727, he had confidants as well as acknowledged and trusted followers. In the sense of being concerned about the way his science would be used, Newton was a Newtonian. Both before and after his fame set in, Newton was a sharp adversary and a demanding friend. As he offhandedly put it, "I am safest in people that are afraid of me." Over the years he made sure, wherever possible, that his followers got the top university positions in the kingdom (even in the Dutch Republic), thus prompting one biographer to describe Newton as the autocrat of science. The title "Newtonian" belongs to the men whom Newton trusted enough to translate his writings or to preach from the pulpits his theories about the power and efficacy of God. He also entrusted them to fight his battles with critics at home and abroad, as Samuel Clarke did in the now famous correspondence with the German philosopher Leibniz, discussed in Part 1. Indeed, Newton could be tyrannical in the loyalty he demanded and the efforts he made to institutionalize his science.

None of these personal involvements in the creation of Newtonianism on the part of its namesake accounts, however, for the different and complex appropriations of Newton's work that occurred in the eighteenth century. Those appropriations began as early as the 1690s, and from then on the first Newtonians set the tone and direction of Newtonianism for much of the eighteenth century. [43] Whether Church of England clergymen like Bentley with an interest in natural philosophy and science, or experimenters like Desaguliers with practical and secular interests, all the early Newtonians had a stake in the established order in church and state. None wished ever to see a return to the turmoil and instability of the previous revolutions. They used arguments drawn from science to oppose political radicals and republicans in the pro-Revolution Whig party; they distanced themselves from the abiding affection that some in the oppositional Tory party still retained for the exiled Catholic king, James II, or his children. Neither Jacobites (supporters of James II) nor secular-minded Whigs, the first Newtonians were supporters of the Protestant succession, staunch Anglicans, reconciled to king-in-parliament as the form of governance; in short, skilled practitioners forging a "middle way" between the extremes of their day. In time, by the 1720s, they would gradually become identified with the Whig party and its ruling oligarchy. In Scotland being Anglican generally meant being Tory, yet among Newtonians of Scottish origin, although sometimes Tories, their support for the Hanoverian and Protestant succession (their anti-Jacobitism) defined their political links to the other English Newtonians.

From this moderate posture and well into the 1740s the first Newtonians waged a cultural war with secular republicans of the "left" and Jacobites of the "right." As one republican, Anthony Collins, put it, the Newtonians "[think of] having caught me at an advantage now that the dispute turns upon points of Mathematicks and Natural Philosophy."[1] Undeterred, republicans like Collins's friend John Toland took up Newton's science and tried to use it to support an entirely materialist explanation of human affairs, of change in history and the origin of government. Just as Newton had feared (p. 21), they used his notion of force to assert the independence of matter, its ability to move by forces now said by republicans to be inherent in it. Despite what the Newtonians preached about the relationship between science and religion, there were many ways of interpreting the meaning of the new mechanical science.

Less materialist republicans may not have gone so far as to say that force was inherent in matter, but they took up the epistemology that rationalized the new experimental interventions and observations of nature. They said that everything in the mind came from sense data, hence that there was no inherent moral sense or image of God carved in every human heart. People learn most about the world through their own observations and not through memorized pieties. Such ideas on psychology or epistemology derived from the philosopher John Locke, who also supported Newtonian science and actively sought to promote it.

But both Locke and Newton would be read in heretical or materialistic ways—much to their respective discomfort. So alarmed were churchmen by the way in which science and the empirical method could be used that they confronted Locke with guilt-by-association arguments: "Sir, you are highly considered and much quoted etc. by the Socinians, deists, atheists, and the bold spirits of this country. You do not approve them, but you ought therefore to disapprove them." [46:42] Locke never said a public word against the republicans, some of whom were allies of his own Whig party. He left the polemics up to the Newtonians and other pulpit orators. A few of them, after all, had access to the greatest scientific mind of the age, and Newton was adamantly opposed to the materialists. No one thought to question the reasons for his public silence.

The first Newtonians did the job that Locke was accused of not doing and that Newton was too fearful of controversy to do himself. They were good polemicists in print and pulpit. Perhaps their private contact with the master scientist and his ferocious piety, his likes and dislikes, had made them that way. Newton had seen to it that his followers were extremely well placed to attack any and all who threatened the liberal Anglican "middle way" between the extremes of republicanism and Jacobitism. And so placed, the Newtonians (with Newton's assistance) articulated a new and profoundly different Protestant temperament. Seventeenth-century English Protestants had long believed that God revealed himself both in His work, in the book of nature, and in His Word, revealed in the Bible. And the *Principia* said more about

His work than had any other book ever published. Thus the stage was set for a significantly new rendering of Protestant religiosity.

Early in the 1690s the young Anglican clergyman Richard Bentley approached Newton to assist in the work of the church. Given what we have now learned about Newton's deep religiosity, he was an obvious choice. The greatest scientist of his age was also committed to the maintenance of Christian and Protestant hegemony in England. That hegemony resided in the Church of England, established then (and now) legally as the official religious body of the state. But as a result of the Revolution of 1688–89 its authority had to be rethought. Bentley and other young and ambitious Anglican clergymen made their careers by inventing a new version of Protestantism based upon the new science and intended to address the challenge of their post-revolutionary situation.

Newton helped Bentley apply the *Principia* to explain the working in the universe of divine Providence. With enthusiasm he wrote to Bentley in 1692: "When I wrote my treatise upon our Systeme [of the world] I had an eye upon such Principles as might work with considering men for the belief of a Deity & nothing can rejoyce me more than to find it useful for that purpose" (see p. 57). Earlier Newton had sent Bentley directions on how to read the *Principia*. Bentley probably never mastered its mathematics, but from the pulpit he could give a credible account of the laws it contained and how they demonstrated the providential hand at work in the universe. The laws demonstrated design and order, thus confirming God's existence. God's work more than His word began to ensure the Church's privileged place. After the Revolution it could no longer censor or prosecute, but now it could use Newtonian science to persuade.

Richard Bentley achieved fame when in the early 1690s he mounted the London pulpits of St. Mary-le-Bow and then St. Martin-in-the-Fields, the church that now casts its shadow onto Trafalgar Square, and explained to the congregation that universal gravitation operating on all the bodies in the heavens proves incontrovertibly the power of God in the universe. Bentley gave his famous set of sermons—the title tells their purpose, *The Folly and Unreasonableness of Atheism* (1692)—in the lectureship endowed by the great scientist Robert Boyle. Bentley's science-inspired piety would have pleased Boyle, and it certainly pleased the hierarchy of the established Anglican church. Bentley, and the other Boyle lecturers who followed him, established what came to be known as the "holy alliance" between the leadership of the Anglican church and Newtonian science. [22] The alliance lasted for much of the eighteenth century, and Cambridge University became its spiritual base. At the essence of this new Protestantism lay an unprecedented explanation for the validity of religious belief; now it was to be based upon evidence drawn from science.

The reorientation of Anglican belief toward a partnership with science, and away from biblical or inherited doctrine, occurred most dramatically after the Revolution of 1688–89. Newton's law-bound universe fitted perfectly with the new political order. Through revolution and parliamentary fiat the country had repudiated absolute monarchy—a move that Newton himself had heartily approved.[2] Yet the gentry and aristocracy wished to see no widening of the political circle. It was necessary to allow all those men who owned income-bearing property to have the vote (about one-fifth of the male population), but never should they be given a share in actual power. Human politics, and not divine will, had made the Revolution of 1688–89; and manmade, although deeply oligarchic, political institutions now governed the country. That stable order lasted for much of the eighteenth century, although it was periodically and ferociously challenged by radicals, reformers, republicans, and eventually by the emerging prosperity of the middling classes.

The task of clergymen like Bentley became to find explanations for order, hierarchy, and human authority that made lawfulness seem to be in the very nature of the universe. In his Boyle Lectures, Bentley explained that nothing could have fitted the physical universe together "except mutual gravitation." Only God, "an immaterial living Mind[,] doth inform and actuate . . . the Frame of the World." Neither chance, nor fortune, nor the random collision of atoms governs any aspect of the natural or human order. "There can be nothing Next or Second to an omnipotent God." The present "system of Heaven and Earth cannot possibly have subsisted from all Eternity." Universal gravitation could only have been placed in the matter of the created universe by God: "For how is it possible, that the Matter of solid Masses like Earth and Planets and Stars should fly up from their Center . . . and diffuse itself in a chaos?" This inherent order includes the human: "All bodies are truly and physically beautiful under all possible shapes and proportions; [they] . . . are good in their kind, and are fit for their proper uses and ends of their natures."[3]

Bentley's message was clear: God is in His heaven, universal gravitation holds the universe together, and all is right and lawful in the postrevolutionary order. Those who would reform the system of government further in the direction of republicanism and away from oligarchic rule are dangerously misguided. Left to their own devices, the random atoms could never make a meaningful world. As a later follower of Newtonianism, Benjamin Martin, put it: "The Democratic School would make us believe that particles of inert matter . . . could dance into form and order, compose harmonious systems of worlds, establish laws of motion, and be productive of increase, life, sense and soul."[4] Those who would substitute belief in fortune or chance, or the randomness of nature, for God's benevolent design found in nature and intended for society are misguided, impious atheists. Equally misguided are those who would treat Christian "morals . . . not as a collection of divine statutes and ordinances sent us by an express from heaven, but only as useful rules of life, discoverable by plain reason, and agreeable to natural religion."[5] Science guaranteed access to heaven's statutes and decrees.

Bentley saw his mission in historical terms. When he explained recent history in his sermons he said that in the course of the Protestant Reformation, and later during the seventeenth-century revolution in England, "we departed from the errors of Popery, and . . . we knew too where to stop." In a gloss that conveniently left out the beheading of the king and the turmoil of the 1650s, Bentley said that the English had managed to avoid fanaticism on the one hand and libertinism on the other. With Providence so clearly visible, only the deists and atheists make reason their sole authority. It is the task of the clergy "to bring over all our adversaries to the truth and power of Religion." And not least, "He who sincerely loves God, also loves Priests."[6]

Such clerically invented arguments for order involved an elaboration upon the originally classical and theological "argument from design," that is, that creation points to, and proves the existence of, the Designer. Now the laws of Newtonian science were said to point to the Designer, and the harmony and order of the macrocosm in turn promised harmony and order in the moral and human order. From the moment Bentley gave his sermon until well into the nineteenth century, design arguments were common in British and American Protestant thought. They owed a great deal to Newtonian science. [25]

The Bentley-Newton relationship is the first and earliest example of Newton's link to the Newtonians, and the relationship served a religious purpose for both sides of the "holy alliance." Arguments drawn from Newtonian science were then used throughout the eighteenth century to bolster belief in Christianity and hence—or at least so the Anglican clergy hoped—belief in the Church of England. It in turn supported the state, and increasingly the leadership of the Church supported the Whig oligarchy that ruled through the state's bureaucracy and through Parliament. And Cambridge became the training ground for clergymen of Whiggish persuasion and Newtonian disposition.

One could be a Tory and a Newtonian, but the species became rare.

Offered by protégés of Bentley like Roger Cotes and Robert Smith, scientific instruction at Cambridge grew to be dominated by Newtonian optics and mechanics. The young Samuel Clarke even edited Cartesian textbooks in such a way as to make them useful for Newtonian instruction. Then in his notes to the *Principia* Cotes made it more readable and understandable—a service its author appears not to have appreciated. Other Cambridge Newtonians taught by Cotes, such as the layman James Jurin (1684–1750), edited still other Cartesian works and subverted their premises with Newtonian footnotes. He also defended Newton and the Newtonians against charges of infidelity made by Tory high-churchmen. Jurin's scientific interests included hydrostatics, and he was capable of doing experiments before the Royal Society to illustrate Newton's theory of force. As a Cambridge tutor, another of Newton's converts, Richard Laughton, drew up a sheet of questions for the use of the schools based upon the Mathematical Newtonian Philosophy. He also urged that the university actively train military engineers, and in his heart—like Newton and Clarke—Laughton probably was an antitrinitarian. The student notebook of William Stukeley (1687–1765), whom we will meet again as a Freemason with industrial interests (p. 103), shows the Cambridge curriculum to have been saturated with the writings of Newton and Boyle. From the universities Newtonianism went on to permeate the schools, in particular the academies run by non-Anglican Dissenters. In them applied Newtonianism became a vital element in a cultural matrix that paved the way for British industrialization. The culture associated with the Dissenters was transferable and exportable, but one element in it remained constant: the emphasis placed upon mechanical knowledge and practices.

Rational Mechanics

Newton's first readers extracted many secular uses and meanings from his science. Indeed, the earliest Continental reception of the *Principia* ignored religion and focused instead on mechanics. In the first two books, this interpreter explained, Newton gives "the general rules of natural mechanics, that is to say the effects, causes and degrees of weight, lightness, elastic force, the resistance of fluids, and the powers called attractive and impulsive." The statement first appeared in a French-language journal of 1688 (*Bibliothèque universelle*, vol. 8, 438–39) and was probably written by John Locke. It set the tone for much of the British eighteenth-century response to the *Principia*.

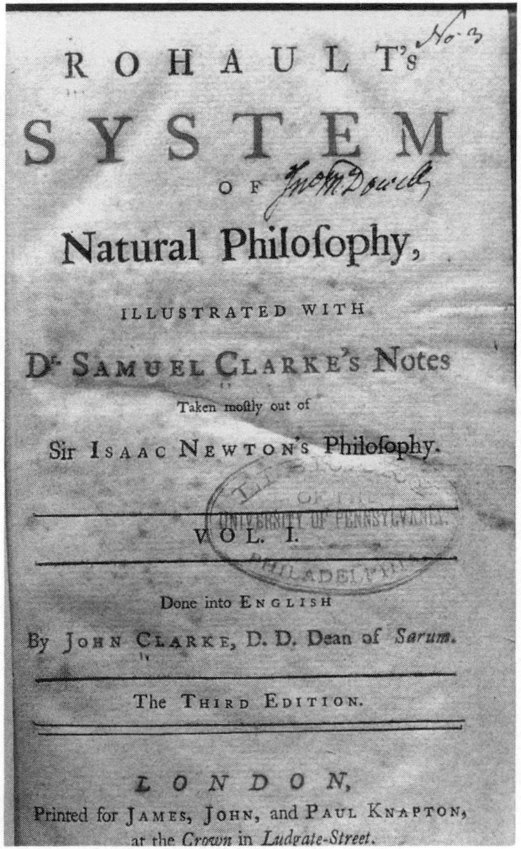

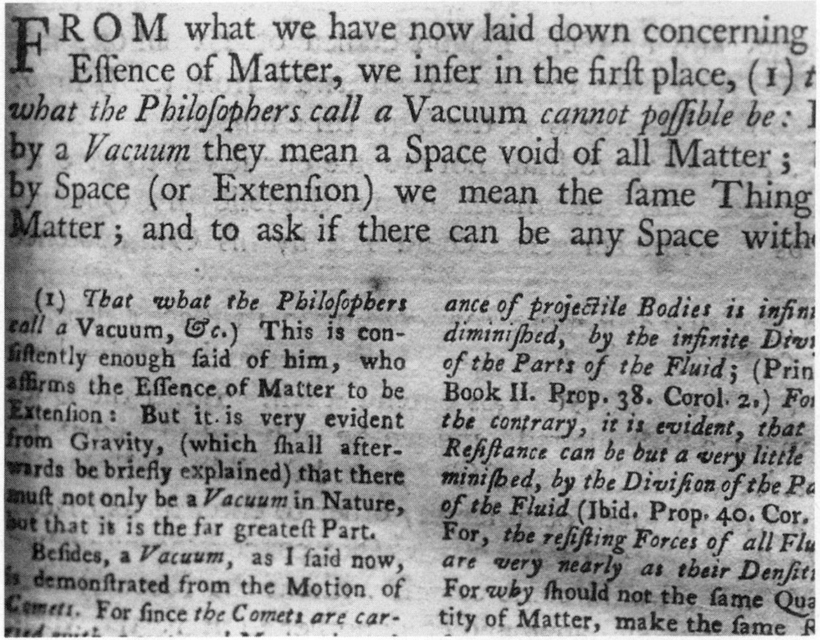

FROM what we have now laid down concerning Effence of Matter, we infer in the firft place, (1) that what the Philofophers call a Vacuum cannot poffible be: by a Vacuum they mean a Space void of all Matter; by Space (or Extenfion) we mean the fame Thing Matter; and to ask if there can be any Space with

(1) *That what the Philofophers call a Vacuum, &c.*) This is confiftently enough faid of him, who affirms the Effence of Matter to be Extenfion: But it is very evident from Gravity, (which fhall afterwards be briefly explained) that there cauft not only be a *Vacuum* in Nature, but that it is the far greateft Part.

Befides, a *Vacuum*, as I faid now, is demonftrated from the Motion of *Comets*. For fince *the Comets are car-*

ance of projectile Bodies is infin diminifhed, by the infinite Divi of the Parts of the Fluid; (Prin Book II. Prop. 38. Corol. 2.) Fo the contrary, it is evident, that Refiftance can be but a very little minifhed, by the Divifion of the Pa of the Fluid (Ibid. Prop. 40. Cor. For, the refifting Forces of all Flu are very nearly as their Denfit For why fhould not the fame Qua tity of Matter, make the fame R

ROHAULT'S SYSTEM OF Natural Philofophy,

ILLUSTRATED WITH

Dr SAMUEL CLARKE's Notes

Taken moftly out of

Sir ISAAC NEWTON'S Philofophy.

VOL. I.

Done into ENGLISH

By JOHN CLARKE, D.D. Dean of Sarum.

The THIRD EDITION.

LONDON,

Printed for JAMES, JOHN, and PAUL KNAPTON, at the Crown in Ludgate-Street.

FIGURE 2.1 The title page from the English translation of the Cartesian J. Rohault's *System of Natural Philosophy*, and a page from chapter 8 where the author asserts that the vacuum cannot exist and (in footnote 1) Clarke painstakingly explains how, on the basis of the notion of gravity, much of nature is made up of a vacuum. Courtesy of Van Pelt Library, University of Pennsylvania.

The 1688 journal alerts us to a long-term trend in the Newtonian legacy. Throughout the century the *Principia* was seen as essentially a great work in mechanics, which Newton mathematicized and regularized as a science. Mechanics pertained to artifice, to intervention by machinery, to the study of "light and local motion"—as Newton defined it in his 1670–72 Cambridge lectures. Mechanics permitted intervention into the processes of nature, and one of Newton's purposes in the *Principia* had been to lay the foundations of this discipline. We now see the *Principia* as a great work in natural philosophy and mathematics, laying emphasis on its application to the planetary system and the elegance of its philosophical underpinnings. But Newton's contemporaries looked to it for a rational, complete mechanics.

The life of Jean Desaguliers nicely illustrates the mechanical side of the Newtonian legacy. He came to England as a small child with a family in flight against religious persecution for being Protestants in Catholic France. Although his life was then molded to be Anglican and clerical, Desaguliers was far less ideologically driven than Bentley—ideological in the sense of having a career primarily as a polemicist in support of the very contemporary concerns of the Anglican church. In contrast to Bentley, Desaguliers spent little of his time in the pulpit and much of it on technological projects, everything from draining lead mines to figuring the flow of water in city pipes. He got his start in Newtonian mechanics when employed by the Royal Society of London to do its electrical experiments.

In some ways Desaguliers, more than Bentley, was typical of practicing Newtonians of the eighteenth century. They were men interested in the practical and the useful, in application, in commercial life and industrial development. In 1714, during the period of Newton's presidency, Desaguliers was appointed the Curator of Experiments for the Royal Society of London, Britain's main scientific society. Based as a pastor in a London church, Desaguliers moved quickly into the president's inner circle. The great man even became godfather of one of Desaguliers's children. With a fine eye and studied concentration, Desaguliers could use a weight swinging on a string with decreasing speed to illustrate the force of inertia, or he could generate electrical charges by rubbing objects against glass bulbs out of which the air had been taken. Newton had long speculated as to the cause of electricity just as he had thought long and hard about the nature of the aether. He sought to understand the effect, if any, of the atmosphere on the transmission of electrical impulses. Desaguliers rigged up various vacuums and showed that whether in vacuo or in the air, electricity affected bodies in remarkably similar ways. The same was true of bodies when chilled or heated. All of this suggested the existence of a subtle medium through which all force or motion is passed. Newton, who watched Desaguliers's experiments avidly, never made up his mind about the nature of these invisible forces, but Desaguliers's career was launched as an experimenter and Newtonian explicator.

Like all good Newtonians, Desaguliers joined the polemics against Cartesianism, and in the preface of his *Course of Experimental Philosophy* he called it a "Philosophical Romance." By applying geometry to philosophy, Desaguliers explained, Newton routed "this [Cartesian] army of Goths and Vandals in the philosophical world." His *Principia* gave the causes of the motion of the comets and planets, explained the action of the planets and the sun on the moon, and stated the true figure of the earth, a flattened spheroid. His *Opticks* revealed the properties of light, and the speculations made at the end of that text "contain a vast fund of philosophy," an agenda for future research. Desaguliers's task, as he saw it, was to use machines to explain and prove experimentally what Newton had demonstrated mathematically. In adopting this technique Desaguliers was building upon the public lectures and experiments first performed by John Keill at Oxford and Francis Hauksbee (d. 1713) in London.

Desaguliers, and English scientists in general, had available to them an infrastructure of daily artisanal practices in applied mechanics that were superior to those found in many other parts of Europe. While the artisans may not have had the theory behind the operation of levers, pulleys, and the friction of metal on wood, they were superior masons, designers, and metal workers who could make the moving parts to pumps and engines sturdy and true to size. Armed with handbooks like Joseph Moxon's *Mechanick Exercises, or the Doctrine of Handy-Works* (London, 1678) they standardized carpentry, joinery, smithing and sundialing. Even Dutch visitors to England in the 1690s, coming from a country noted for its instrument making and its metal and wood working, thought that houses in London were better built than what they saw at home, and they commented on the skill displayed in everything from furniture design to the casting of metal rollers.[7] Gradually Newtonian mechanics organized all these practices into one coherent, theoretically anchored science, but its rationality was deeply indebted to artisanal practices that in time paved the way for industrial development.

With good machines and devices available for demonstrations and experiments, all the Newtonian commentators liked to begin with first principles illustrated by simple experiments, then work their way up to greater levels of difficulty. As Desaguliers explained in the opening lecture of his *Course* (vol. 1, 4), matter is the same in all bodies, and

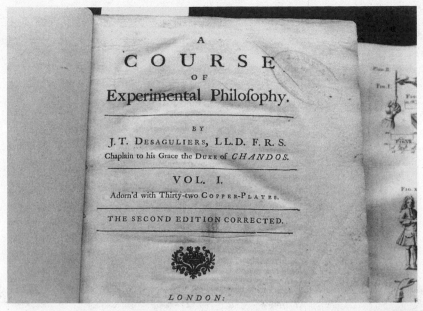

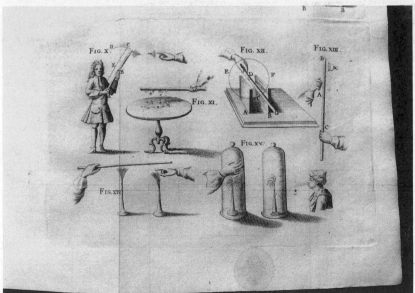

FIGURE 2.2   From Desaguliers's *Course of Experimental Philosophy*, vol. 1, illustrations of electrial phenomena caused by rubbing or friction, which could be performed on any tabletop. Courtesy of Van Pelt Library, University of Pennsylvania.

"the whole variety of bodies, and the different changes that happen to them, entirely depend upon the situation, distance, figure, structure, powers, and cohesion of the parts that compound them." Mercury offers more resistance than water, water more than air, because there is a greater number of particles contained in the same space in the heavier body. The atoms are the constituent parts of all bodies "which the wise and almighty Author of Nature did at first create as the original particles of matter."

On the very same page—we may assume in the very same lecture—where this Protestant piety about God the creator of the atoms appears, Desaguliers illustrates experimentally the vacuity between the atomic particles by emptying the air from a cylinder and allowing a feather and a coin to drop simultaneously, landing (as only will occur in a vacuum) at precisely the same moment. In one experiment Desaguliers had neatly illustrated Boyle's vacuum, the uniformity of matter, and set the stage for explaining that in the heavens "that force which bodies have when they are carried towards each other, which (at equal distances) is always proportionable to their quantity of Matter." The fascinating and compelling quality of Newtonianism lay in the totality of its explanatory power: atoms, small bodies, the earth, the sun, and the planets fit together in one conceptual whole. In the hands of the early Newtonians, Newton's text moved from being a work in philosophy, *The Mathematical Principles of Natural Philosophy*, toward being the foundation for modern science, for experimental inquiry and its application. It became simply the *Principia*. [14]

Assisted by Newtonian mechanics, Desaguliers also had a vision that may be called proto-industrial. He understood the earliest steam engines, and he had studied the application of the laws of motion particularly to the movement of water. He measured frictions and provided tables about them produced by the effects of iron, lead upon wood, brass, copper, and so forth. He demonstrated the impact of weights upon carts and wagons, and he provided yet more tables that showed what forces were required to bend ropes of different diameters, stretched by different weights, around rollers of different sizes. His instrumentation was elaborate and its effects dramatic. He could do more than 50 entertaining experiments with the air pump, which he and his fellow practitioners had vastly improved since, more than 60 years earlier, Boyle's leaky pump first confirmed the existence of the vacuum. Historians used to argue that Newtonianism had little to do with actual practices, applied and potentially industrial. But the research of the last 20 years has demonstrated the practical side of the Newtonian legacy, thus connecting the achievements of the Scientific Revolution of the seventeenth century to the transformations wrought by the

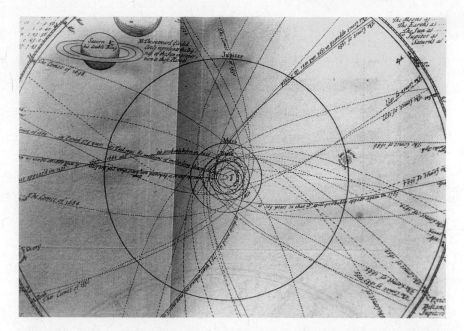

FIGURE 2.3   The Newtonian universe as illustrated in Desaguliers's *Course of Experimental Philosophy.*

Industrial Revolution of the eighteenth.

The connections with the Industrial Revolution can be seen in Desaguliers's lectures and his engineering activities. From tabletop pumps it was a natural step to move on to freestanding, vacuum-driven engines and to caution that "power lost, or misapplied, bad materials, unnecessary frictions, oblique fractions where they should have been perpendicular, animals working in disadvantageous postures; streams of water half lost, back-water returning, whole machines ill put together, etc., have been the cause" of the bad performance of many engines. Desaguliers's pitch aimed to undermine "all the plumbers and mill-wrights" who without proper training in Newtonian mechanics and mathematics had gone into the business of engines.[8] Likewise he had little use for "men of theory" who did not understand the work of bricklayers, masons, carpenters, the strength and coherence of bodies, and the practices of machines in each of their particular parts. The message about machines had an audience. By 1730 there were already over 100 steam engines, named after their inventor Thomas Newcomen, installed in Britain. [17]

Desaguliers was creating a profession for himself, that of the civil engineer. His theoretical sophistication derived from Newton's writings, out of which he extracted principles useful to manufacturing. The 1717 edition of Newton's *Opticks* spoke about the elasticity of fluids. Desaguliers applied the concept to steam, arguing that it maintained its heat and elasticity and could therefore be contained in copper vessels in close contact with other volatile substances in need of heat but too dangerous to place near fire. Through the transfer of steam heat from vessel to vessel even gunpowder could be dried safely. [77] In cooperation with a coppersmith and an instrument maker, Desaguliers presented his device to the Royal Society, and in 1730 he secured a patent for it. Desaguliers did not die a rich man, but it was not from want of trying. In many aspects of his career Desaguliers was more entrepreneur than clergyman, more engineer than simple experimenter.

The Spread of Newtonianism

In lectures given throughout Britain and in the Low Countries (where he spoke in French or Latin), Desaguliers offered something for almost any skilled craftsman, merchant, or self-improving listener. The published version of his lectures, complete with careful and exquisite drawings, could be used for the education of other Newtonian practitioners or for self-education by entrepreneurs whose success depended upon the skill of the engineers whom they hired. A course of a dozen lectures cost about two guineas, the price of dancing instructions with a good master. By virtue of his breadth and clarity, Desaguliers became the most famous of literally hundreds of eighteenth-century Newtonian practitioners who applied the mechanical legacy of the *Principia* to a wide range of practical problems, from the draining of mines and swamps to the building of canals and electrical experimentation.

Courses in all these topics became institutionalized particularly in the academies run by the non-Anglican Dissenters. As late as the 1820s the natural philosophy curriculum found in them closely followed the topics set forth in Desaguliers's two-volume *Course* published in the 1740s.[9] As we shall see shortly, such courses cannot be separated from the cultural history of early industrialization in Britain. Even more so than Cambridge, the Dissenting academies, whether in London or Manchester, led the field in science education, and by the late eighteenth century any self-respecting business family with industrial interests opted first to send their adolescent boys to such an academy, or possibly to Edinburgh University, where medical and scientific education was among the best to be found in Europe. In these schools of higher education students used mechanical instruments and devices.

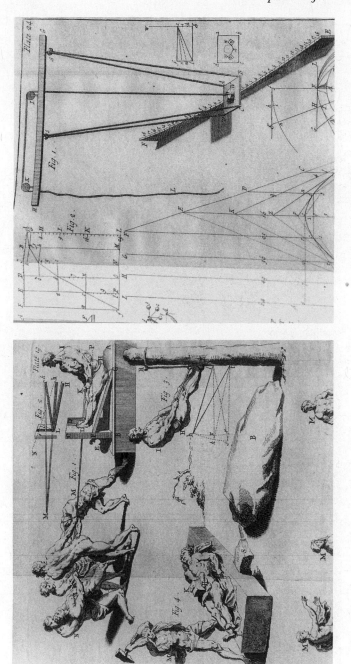

a

b

FIGURE 2.4 All illustrations taken from Desaguliers's *Course of Experimental Philosophy*, these pictures first explained how human strength must struggle against gravity and inertia (2.4a), then illustrated how these forces can be measured (2.4b), went on in subsequent lectures to discuss the movement of water (2.4c) and the removal of coal from mines (2.4d), and ended with contemporary steam engines and water wheels (2.4e, f). Courtesy of Van Pelt Library, University of Pennsylvania.

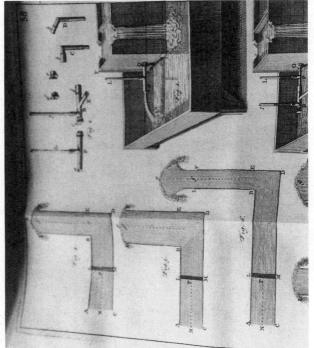

d

c

Plate XXVII.

front. p. 442.

A WATER MILL for grinding torn at the BARR Pool, by Y. Alley in Nun-Eaton in Warwickshire.

Plate XXVI.

front p. 442.

f

e

They learned hands-on mechanics, hydrostatics, pneumatics, or optics, or they could begin medical education that had been shaped by experimental approaches indebted to Newtonian science.

Part of Desaguliers's success as a lecturer, and indeed the efficacy of Newtonian demonstration in general, depended upon the quality of instrumentation available in London, the English provinces, and then slowly in the American colonies. In the seventeenth century the Dutch had been the greatest instrument makers, but gradually the art shifted across the English Channel. To this day eighteenth-century instruments can be seen in museums or in the collections of older British and American universities from Oxford to Harvard, Yale, Columbia, Princeton, and William and Mary in Virginia. [37] At the College of Philadelphia, later the University of Pennsylvania, students were required to spend 40 percent of their time on scientific subjects. Air pumps, telescopes, globes, thermometers, orreries—all were part of standard classroom experience. [3] Orreries re-created the Newtonian universe, complete with moving parts that could illustrate the positions of the planets by the turn of a handle that set this miniature universe into harmonious motion. For the wealthy, orreries, exquisitely crafted in brass and wood, could be purchased as objets d'art, thus bringing the Newtonian universe into the very microcosm of domestic life. While engineering became an exclusively male preserve, scientific knowledge slowly entered the mental universe of highly literate women and girls.

Most public lecturing quickly came to include electrical experimentation. Already of considerable interest to Newton and to the other Fellows of the Royal Society, the phenomenon of electricity also just delighted and awed observers. In Britain Desaguliers was the most important electrical experimenter of his generation (in the American colonies it was his contemporary Benjamin Franklin). Desaguliers studied charges, conduction, attraction and repulsion of electrical particles, the effects of dryness and moisture. He distinguished between bodies that could be electrified and those that appeared naturally insulated from electrification except when suspended. He did not understand insulation, nor did many of his fellow practitioners in what could be a dangerous pastime. Following in his wake, by the second half of the century British electrical experimenters were using charges even in medical treatments of everything from lumbago to the gout. [74] The theoretical foundations for all these efforts to use electricity had been laid down in the queries to the *Opticks* (1723) where Newton had speculated on the properties of the aether, the miniscule effluvia that seemed to infuse nature and also seemed to be controlled by the universal principles of attraction and repulsion.

But Desaguliers was also more than an electrical experimenter and traveling lecturer. His linguistic and mathematical skills permitted him a wide variety of contacts abroad, and he translated into English from Latin a mathematical work by the Dutch Newtonian Willem s'Gravesande (pronounced skrave-san-da). S'Gravesande's *Mathematical Elements of Natural Philosophy, confirm'd by Experiments; or, An Introduction to Sir Isaac Newton's Philosophy* was one of the important eighteenth-century text-books that made Newton's mathematics accessible and that also anchored the philosophical foundations of Western science on "evidence and stability which have put mathematics out of the reach of error and contention."[10] For its Dutch author, as for its English translator, this was an ethical and religious enterprise, intended, as its preface proclaimed, to show "that God is good, and that this appears also by mathematical demonstrations." God's goodness lay in the order and regularity of nature, now explicated mathematically. By the early eighteenth century the sensibility of a Dutch Newtonian like s'Gravesande was remarkably similar to that of other Newtonian Protestants. [81] The new science-inspired Protestantism spread throughout Continental Europe.

Like public lectures in England, s'Gravesande's much more technical account of Newtonianism, originally given in Dutch, also began with the atoms and with density and elasticity, then went on to calculate mathematically the force acting upon an inclined body: "Let the surface of the glass be AB; a particle C; this tends towards the glass in the line CD, perpendicular to the surface; it also tends towards the Point e . . . ." From a simple mathematical demonstration of attraction and repulsion in the straight-line motion of a drop of water on a glass surface, the text moves to a discussion of motion in general, the force of gravity, the vacuum experiment to illustrate the effects of gravitational force on bodies without resistance, then on to levers, weights, and pulleys. To these experiments "mechanical arithmetick" is added, and arithmetic along with geometry is used to calculate the power of machines.

Eventually Newtonianism subjected all nature to mathematical treatment: "The water that runs by its own gravity, in a channel open above, is called a river." So began s'Gravesande's mathematical explication of hydrostatics, the mechanical practices so necessary for canal building and the control of water pressure. By the end of his text the mathematician, although by a more technical and theoretical route, has also become a civil engineer.

As part of his professorship in Leiden s'Gravesande acted as a consultant on water projects in the Dutch Republic. In his youth he was also the member of a secret club that put out the first journal published

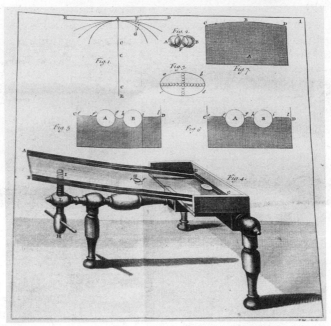

FIGURE 2.5   The illustration for the simple motion of a flat body on an inclined plane, from W. J. s'Gravesande's *Mathematical Elements.* Courtesy of Van Pelt Library, University of Pennsylvania.

in French (but outside of the control of French censors) that disseminated Newtonianism to French-speaking Europeans. The *Journal littéraire* emanated from the Dutch Republic and it flourished for only about a decade; it was over by the mid-1720s. Yet its influence spread deep into Europe and its colonies; south of the Alps Italian intellectuals like Celestino Galiani acted as corresponding editors. Early in the 1720s a European merchant in the distant Dutch slave colony of Surinam could write to one of s'Gravesande's co-editors and tease him by saying, "I will abjure my Cartesianism if you will adjure your Newtonianism." Desaguliers, s'Gravesande, Hauksbee, and a half dozen other early Newtonians, at work in lecture rooms, print shops, and editorial boards, had succeeded in spreading their master's science along with the Newtonian version of natural religion wherever European trade and empire had also penetrated. Newton's legacy became distinctively Western and undoubtedly contributed to the sense of superiority that sanctioned European wealth and overlooked the misery of others.

Dutch pupils of s'Gravesande took his ideas to other universities and academies in the Republic, while the French-language press sent them far and wide. French was the international language of the age. Even English students learned Newtonianism from s'Gravesande in Leiden and then brought the knowledge back to Britain. In addition, the greatest doctor of the age and a professor of medicine at Leiden, Hermann Boerhaave, also took up Newtonian principles to describe the life of the body as dominated now by the principles of attraction and gravitation. As a practicing doctor Boerhaave was impressed by the experimental, practical side of the Newtonian style; it further reinforced his insistence upon hands-on observation and examination. In Leiden he became famous as the doctor who cared enough to venture into the homes of the sick and poor, to study disease as it was lived.

Boerhaave was quite simply the most important medical practioner of the age, and his practices became the norm in most medical schools by the middle of the century. [88] The greatness of medicine at Edinburgh derived from the years that its faculty had spent as students with Boerhaave in Leiden. By the middle of the eighteenth century, the extent and range of European domination enjoyed by Newtonian science, both practical and conceptual, had become wider and more sophisticated than even Newton—master experimenter as well as conceptualizer of space and time—could ever have imagined.

Thus no longer clerical, many Newtonians became worldly and cosmopolitan experimenters and showmen. S'Gravesande never took clerical orders, while in the case of Desaguliers being a clergyman was even something of a hobby. Although they paid his salary, preaching and caring for souls did not much interest him. One of his aristocratic patrons, for whom Desaguliers acted as an engineering consultant on his estates, became so incensed by Desaguliers's neglect for his parish that he sent this rebuke: "The inhabitants . . . have been forced to go a begging to other minsters to bury their dead. This is very shameful neglect." [76:219] When the great contemporary satirist William Hogarth did a popular engraving called *The Sleeping Congregation,* the preacher giving the sermon was widely regarded as none other than Jean Desaguliers.

In a bad poem (it is at least for us interesting) Desaguliers used Newtonian imagery to celebrate the coronation of George II (1728). *The Newtonian System of the World, the Best Model of Government* (p. 17) summed up his understanding of the new postrevolutionary relationship between English government and Newtonian science:

> What made the Planets in such Order move,
> He said, was Harmony and mutual Love.
> The Musick of his Spheres did represent
> That ancient Harmony of Government

Although Newton had little use for poetry, it was nevertheless a bless-ing that he died the year before the poem was published. Newton did not imagine that God had placed universal gravitation in the universe in order to glorify the institutions of mere mortals, even of kings.

By the end of the first quarter of the eighteenth century a far more secular understanding of history and time had come to prevail. Ironi-cally Newton's science had helped to usher in an expansive and opti-mistic age in which he would have been a relative stranger. When Voltaire came to London in the mid-1720s he was stunned by the spread of Newtonian ideas, by the relative freedom in publishing and politics, and by the ease with which the literate classes seemed to socialize with one another. By ease he meant that the English aristocracy willingly fraternized with their lessers. Indeed, in Voltaire's mind all three phe-nomena were of a piece, and he told the world about it in his *Letters on the English Nation* first published in French in 1733. England was modern because Newton's science was better, its cultural life thrived on freedom, and its aristocracy respected men of learning and com-merce. With wit and sarcasm Voltaire gave voice to the Anglophilia of the age, and his book was the single most important work of the cen-tury that introduced Newton's name and his principal achievements to Western readers. Voltaire added to its impact by publishing his own, rather technical exposition, *The Elements of Newton's Philosophy* (French edition pirated in 1738; Voltaire's approved edition, 1741).

With supporters like Voltaire (who was a deist), perhaps we can better understand why by the late-eighteenth-century Methodist ministers raved against Newton, calling him an ignorant pretender. Even John Wesley himself, the founder of Methodism, was ambivalent about the mean-ing of Newtonian science. For people who wanted Protestantism to be biblically based and from the heart, not the head, the Newtonian ver-sion of Protestantism seemed effete, a sop to the decadent mores of the rich or wellborn. Late in the century William Blake, the great radical, religious mystic, and prophet against empire, saw Newton as a symbol of human beings enslaved to material things, the creatures of science rather than the creators of new worlds. Science had become irretriev-ably associated with indifference, if not hostility, toward religion. The Whig oligarchy that so impressed Voltaire had helped to make liberal Christianity seem like a fig leaf to cover their greed and corruption.

Other radical critics of the age took up science not to encourage complacency but to undermine centuries of traditional belief and the power of the clergy. Man the machine, rather than man the sinner, became a metaphor of choice for atheists and materialists. Trained in science, they were capable of writing books with titles like *L'Homme machine* (1748) wherein all human action and emotion were seen to result from the movement of nerves and muscles. The metaphor horrified Newtonian as well as Christian critics, and it did not help that the author of *Man the Machine,* Julien Offray de La Mettrie, had been a student of medicine with Boerhaave in Leiden. [86] The Dutch publisher of the book had to write a book himself explaining that he did not endorse the ideas he published but did believe in freedom of the press. [82] When the English republicans like John Toland took Newtonian force and attached it to matter, just as Newton feared would happen, they invented a new and virulent form of materialism that influenced European and American thinkers as important as the French philosophes, Denis Diderot, the Baron d'Holbach, and Thomas Paine. The radical side of the European Enlightenment owed as much to Newtonian natural philosophy as did the moderate, more Christian side of this new cultural movement. Inspired by nature as revealed through science, radicals like Paine called for a philosophical religion that would encourage democracy.[11] None of the early Newtonians would have approved.

Even less radically materialist or pantheist philosophers imagined that mechanics stood as a model for making all human systems into sciences. Western thinkers had always appealed to nature for imagined laws to govern behavior; now they argued that morality itself might be reduced to "a system as well connected, as those of geometry, for example, or mechanics, and founded on as certain principles."[12] Despite the association with atheism and materialism, some theologians found in the machine the metaphor for divine design, a symmetry found in watches, hydraulic machines, wheels, even in the parts of the human body. At the end of the century, when searching for a new science of society, William Paley triumphantly proclaimed that "in the works of nature we trace mechanism; and this alone proves contrivance"—by the Divine Artificer. [26] Paley had been directly inspired by the industrial machinery that by the 1790s was increasingly visible in the English countryside. The popularity of the design argument was as old as Richard Bentley, indeed older, but now late in the eighteenth century industrial machinery gave it a new lease on life.

Thanks to the publicity generated decades earlier by Desaguliers and others, by midcentury mechanics as well as mechanisms of every size and description had become everywhere fashionable, and automata fascinated both the great and the lowly throughout northern and western Europe. Imagine an automaton musician playing a flute, a metal duck eating, drinking, and defecating (then picking its feathers), and another machine dressed like a dancing shepherd and playing a tabor and pipe. "I forgot to tell you, that the Duck . . . makes a gurgling noise like a real living Duck." Or so Desaguliers's translation of the lectures of the French mechanist Vaucanson told its readers. "The inspection of the machine will better shew that nature has been justly imitated," right down to the wings. The thing did not fly, but it did not take great imagination to ask, Why not? Throughout Europe the machines were presented along with learned but simple treatises that explained the underlying principles: "All these vibrations of the mouth may be performed . . . because the wind . . . must always be so regulated . . . it will have vibrations equal to those that are produc'd in the middle of the note where the sound increases in force, because it will be communicated to a greater quantity of air."[13] The fad for automata had nothing directly to do with Newtonianism, but the fashion assisted in the assimilation of Newtonian understandings of both the universe and local motion. It also helped to naturalize machinery from simple button makers to steam engines.

By the middle of the eighteenth century European children of some affluence were being taught to think mechanically while their parents, first men and then gradually women, were attending lectures in Newtonianism. [70] This European trend had started, however, in England a full generation earlier. Even the philosopher John Locke, as early as the 1690s, had written a simple treatise on natural philosophy for one of the wealthy pupils in his charge. There is some evidence to suggest that Newton gave him a hand with it. Decades later the manuscript treatise for children was partially copied in a popular science book, *The Philosophy of Tops and Balls,* published by one "Tom Telescope." [2] The only surviving diary written by an eighteenth-century adolescent tells us that in the Dutch Republic of the 1790s a boy of affluent parents routinely attended scientific lectures and read books about science and natural religion.[14]

Decades earlier in Paris, the equivalent of the weekly wage of a young worker got someone into a concert of flute music played by an automaton. In 1738 thousands paid to hear it. Vaucanson, the flutist's creator, made his money and quickly turned his attention to industrial machinery. In France, as elsewhere, applied mechanics was the route to profit and social prestige. Yet in France public lectures were nowhere near as popular and hence lucrative as they were in England. In Paris the abbé Nollet imitated Desaguliers and made a living with his lectures, but the professors at the Sorbonne were hostile to him and even put pressure on the authorities to close him down.

Nollet's textbook in mechanics was far more successful than his Paris lectures, and it was widely used in French engineering schools by the middle of the century. By then most French engineers knew Newtonian mechanics, but almost to a man they were employed by the state, as befit their aristocratic origins. As we shall see shortly, the strength of French science lay not in its widespread availability but in mathematical originality as well as in theoretical innovations. These took Newtonian science in new and exciting directions. But on either side of the English Channel, being au courant with mechanical science, however superficially, signaled appreciation for innovation and modernity. No wonder that Queen Caroline decorated her garden with busts of Newton, Samuel Clarke, and Robert Boyle, among others. The first scientific society for, and organized by, women was established in 1785 in the Dutch city of Middelburg and used Nollet's textbook as a teaching manual.

The Newtonians of Desaguliers's generation and beyond offered, as Benjamin Martin said, nothing less than "utility, pleasure, and happiness" to their age. Showmen with lofty as well as practical messages, sometimes self-educated, found careers as itinerant lecturers. [24] They brought the Enlightenment out of the drawing rooms of the elite and to the middling classes, proclaiming science the key to human progress. Through mechanics in particular, Martin said, "we have here opened to our minds the wondrous laboratory of nature, and the stupendious processes therein carrying on, unheeded and unthought of by the vulgar." [54] Even earthquakes, volcanoes, and comets, although dangerous and destructive, were no longer "secrets in the school of natural philosophy." In addition, it was claimed, every mechanic and manual trade could be improved by Newton's science. Now applied science promised self-improvement, profit, and status.

Mechanical knowledge became one key to social mobility, and lecturers such as James Ferguson went up and down the English countryside telling of their life histories and the prosperity that came their way through scientific knowledge. [55] There was truth in the boast that applied science was one way up to a respectable living. Even in Newton's lifetime Francis Hauksbee, who had made some of the earliest electrical experiments seen at the Royal Society, had had no formal education and was a draper turned instrument maker and physics lecturer. In the early eighteenth century his writings were even known in Italy, while in the 1780s a Dutch instrument maker in Zeeland built a planetarium using a translation of Ferguson's lectures. Benjamin Worster, a friend of Desaguliers, lectured in bookshops, Masonic lodges, and academies for craftsmen and reached artisans, merchants, and tradesmen, any one of whom could have understood his clear and simple rendering of natural philosophy, mechanics, the laws of motion, and hydrostatics.

But the spread of science and opportunities for scientists, however

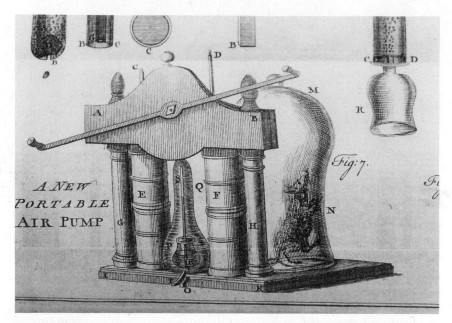

FIGURE 2.6   Benjamin Martin used a portable, miniature air pump to
illustrate the vacuum. It is not clear if he released the mouse from its
airless vacuum before, or after, it expired. From *A Panegyrick on the
Newtonian Philosophy*. Courtesy of Van Pelt Library, University of
Pennsylvania Library.

humble their birth, differed from country to country. By the mid-
eighteenth century the number of working scientists in England equaled
that found in France, a country with nearly twice the population. [23]
Even in the highly literate Dutch Republic a lack of interest in mechanical application—but not in general science—hampered the em-
ployment opportunities of practicing engineers. In no place were the
vast majority of practitioners original thinkers or innovators. But at no
time in history has innovation ever been anything other than exceptional.
   In England the Newtonian lecturers offered many enticements: men
and women need only read, or better still, go to experimental lectures, and with knowledge of science would come national, if not personal, superiority. Despite his own humble origins Benjamin Martin
smugly lectured on the contempt with which the English held the
Chinese, the Tartars, the Indians, etc., for their lack of erudition and
the study of the sciences. British national superiority was explicitly linked
to Protestantism and scientific superiority, and Newton's science em-
bodied both of them. By the late eighteenth century a British nationalist

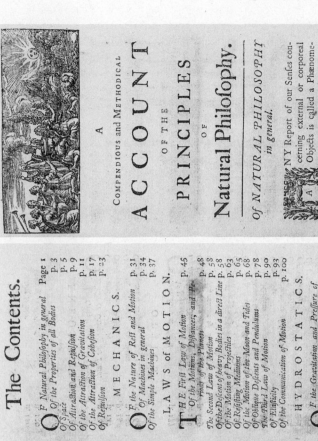

FIGURE 2.7 The title page, a partial list of contents, and the opening of one of Worster's lectures show how clear and simple Newtonian mechanics could be made. Courtesy of Van Pelt Library, University of Pennsylvania.

ideology had coalesced, and scientific and technological superiority was one vital piece in its construction. Throughout the English-speaking world, but especially in the recently independent United States, Anglo-American progress in the eighteenth century came to be seen in retrospect as scientific, political and moral. If one single man could be credited with exemplifying the progress of the eighteenth century, Isaac Newton became the candidate of choice.[15]

Throughout the Western world Newtonianism emerged in the hands of its popularizers as the first body of scientific knowledge also specifically addressed to women. *The Ladies' Diary* printed tables showing the diameter of pump barrels and steam cylinders required in order to pump a given quantity of water from a particular depth. In the 1740s Eliza Haywood's journal for women, *The Female Spectator*, claimed in volume 3 that any woman could make herself scientifically literate by a summer of intense reading. It further urged women to take magnifying glasses on their walks and to report back to the Royal Society any interesting discoveries. Whole treatises explicating Newtonian astronomy were done as dialogues between a genteel brother and his sister.[16] To be sure, he was always the mentor; but however gained, knowledge gained also represents empowerment. For some women scientific education may have been more accessible than the traditional classical education associated with the clergy and the universities.

Far more germane to the story of women and Newtonianism stands the work of Madame du Châtelet, whose *Institutions of Physics* (1740) made her one of a handful of important Continental Newtonians. She also did what remains to this day the only French translation of the *Principia*. Along with Voltaire, her lover, Madame du Châtelet entered directly into controversies concerning the exact nature of the forces that communicate motion. Her mathematical assistant, Samuel König, studied with the best Newtonian mathematician of the age, Maupertuis (1698–1759), and went on to become one of the first historians of the Scientific Revolution. His lectures to an elite audience in The Hague have never been published, but while still respectful of Cartesianism, they tell of the gradual triumph of Newtonian ideas in Continental Europe. [41] Du Châtelet and the Italian scientist Laura Bassi rank as the most important women scientists of their age. Late in the century the women's scientific society in Middelburg, composed of some 40 elite women in a city of 15,000, gathered to learn science and cited Madame du Châtelet, along with Newton and Descartes, as inspiration. Their teacher was Daniël Radermacher, a Voltairean and freemason who was devoted to scientific learning. Inspired by the principles of the Enlightenment, the society flourished into the 1880s.

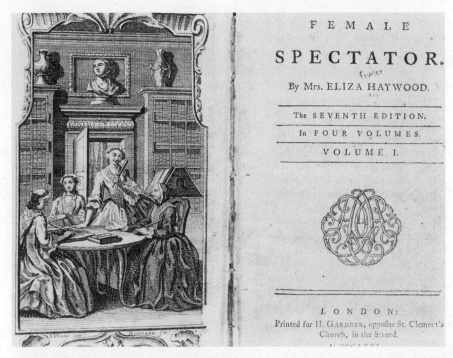

FIGURE 2.8   *The Female Spectator*. Courtesy of Van Pelt Library, University of Pennsylvania.

Also an early Newtonian, the Italian professor Laura Bassi was the first woman to be offered an official teaching position in any university in Europe. First she taught Newton's *Opticks* and experimental physics in her own home, and then she risked controversy and took Newtonianism into the university. In 1749 she presented a dissertation on gravity, and in the 1760s she made an original contribution to the problem of refrangibility. In collaboration with her husband, Bassi made Bologna into a major center for Newtonian instruction and experimentation. [18]

The approach taken by *The Ladies' Diary* vastly underestimated the effort involved in mastering the new science, but Newtonian feminists such as Francesco Algarotti sought to facilitate women's mastery. Directly inspired by Châtelet and Bassi, Algarotti now self-consciously addressed Newtonian science to female readers. Coming out of an Italian context where the new science was locked in combat with the Roman Catholic church and the Inquisition, Algarotti sought to draw women into the fight on the side of science. Yet in the 1740s his book on

Newtonianism for women became a success all across Europe. It was openly anticlerical and hostile to the philosophy taught in Catholic universities.[17] Algarotti also pointedly allied Newtonianism with revolutionary ideas in cultural matters while simultaneously seeking to assure his female readers that there was nothing to fear in science, the progress of which only the "rude and uncultivated" retard. It was a bold, although deeply class-bound, argument, and it won converts among educated women like Lady Mary Ashley Montagu, who was deeply attracted both to Algarotti and to his teachings. Except for rare cases in Italy, all formal European scientific academies excluded women; this did not stop them from being scientific and in some cases specifically Newtonian. To imagine that in the eighteenth century women perceived science as a hostile body of knowledge is to engage in a historical anachronism. Typically the rhetoric and metaphors used to describe nature made "her" feminine and urged her conquest. But most rhetoric and iconography also made Truth to be a woman; only male prejudice hindered easy access to either truth or science.

Late in the century the leading feminist of the era, Mary Wollstonecraft, took courage from science and identified it with progress and enlightenment. Indeed, her own conviction that she was at the end of the age of brute strength as the determinant of power, and hence of male domination, may have been partly inspired by her knowledge of mechanisms and their power. Certainly applied scientific knowledge and conversation were commonplace in the republican and radical circles in which she traveled. Many of the men and women in that circle, such as James and Annie Watt and their son, were at the heart of the application of mechanics, in particular steam, to the industrial process. They were also early supporters of the French Revolution, as was Wollstonecraft.

Wollstonecraft may also have dimly sensed the quiet revolution that was occurring in mechanized manufacturing, although not perhaps in the lives of working women and children drawn increasingly into factory production. [5] For these often illiterate or semiliterate women, the machine metaphor described the actual motions of their hands more than the movement of the Newtonian heavens. They knew very little about the scientific knowledge that gave some of their employers the confidence to buy and deploy the new mechanical engines.

Newtonian science and the culture it inspired captivated many educated women, and only a crude reading of gender divisions would designate either as simply the property of men. Rather the rationalizing and confidence inspired by science now belonged to the literate, and a great divide widened between the folk practices of illiterate women and men, and the ever watchful, trained eyes of those men and women to whom only science-inspired models of nature and society began to make any sense. [80]

Newtonian Variations on a Christian Theme

In an age still dominated by religious convictions and practices, knowledge about nature inevitably required adjustments in the Christian beliefs and rituals surrounding nature's God. In addition, Newton's science rested on certain philosophical assumptions that from the perspective of traditional Christianity were problematic. As we saw in Part 1, most Christian theologians of Newton's early lifetime, both Protestant and Catholic, were Aristotelians. Aristotle's philosophy of nature as transmitted by textbooks, and seldom by a careful reading of the ancient Greek master, was the backbone of most philosophy and theology classes in the schools and universities of early modern Europe.

One of the central assumptions of Aristotelianism, the doctrine of "forms," was, however, simply incompatible with a mechanical, atomistic understanding of bodies in motion, and that understanding lay at the heart of the new science from Galileo to Newton. Similarly, English scholastics of the seventeenth century such as Alexander Ross believed that the "qualities" of bodies were real. In *The Philosophical Touch-stone* of 1645 he said: "Actions have their original from qualities. . . . The hen by her heat, which is a qualitie, prepares the matter of the egge for introduction of the forme of a chick." Qualities were as real as forms (and masculine beings, whether clerical or not, often had greatest access to them).

In conflict with Aristotelian philosophical commitments, the new science had a precarious place in the curriculum of most universities—all dominated by the clergy—and it was an object of deep suspicion among many theologians, Protestant but especially Catholic, and also among university professors. Their penchant for "forms" and "qualities" permitted explanations for a whole host of things that were dear to the heart of the authorities of both church and state. What made a king kingly, when manifestly he looked and frequently acted like any other mere mortal? He was God's anointed, answered the theologians; his soul, the "form" that encased his body, had received a special grace. How could the sacrament of the Eucharist, the centerpiece of the Roman Catholic mass, really contain the body and blood of Christ when, once consecrated by the priest, it continued to look just like ordinary bread and wine? The priest had the power, Catholic theologians taught, to change its "form." Its matter remained the same, bread and wine, but its invisible form changed into the divine. Forms made brute matter into the things we see: wood becomes a table because the crafter has in mind the form "tableness," which can be imparted only by his artifice. Similarly, hens have within them the capacity to receive the form of "chickenness." Likewise heat is a real entity, not—as the new science taught—the result of friction, of particles of matter set in motion by contact with other particles. None of these arguments, essentially of medieval origin, would survive among educated men and women of the eighteenth century who embraced the new science. [72]

By the early eighteenth century in England a new breed of theologian and philosopher had come upon the scene, and these essentially Newtonian thinkers reorganized the way all Christian intellectuals, except the most devout and traditional Calvinists and Roman Catholics, understood the world around them. The intellectual leader of this new generation of churchmen was not Richard Bentley—he was essentially a skilled pulpit orator—but Samuel Clarke. His Boyle Lectures of 1704–05 were far more philosophically rigorous and informed by Newtonian science than Bentley's had been. Clarke had learned the new science from the master himself, and with Newton's assistance he had set himself the task of providing a new philosophical foundation for Christianity. [77]

Clarke's mission was both intellectual and social. He saw that the old Aristotleianism no longer worked; why have "forms" to explain change in the world when mechanical actions and the movement of atoms would serve just as readily? But Clarke also saw that religion was necessary for social stability: "Even the greatest enemies of all religion, who suppose it to be nothing more than a worldly or state-policy, do yet by that very supposition confess thus much concerning it." [43:182] For Clarke the task of shoring up the philosophical foundations of revealed religion was an urgent one, and he gave philosophical depth to the Newtonian version of Christianity.

The urgency arose partly because of the poverty of Aristotleianism, but largely because Clarke perceived that in the heart of Newtonian science lurked an entirely naturalistic, secular way of explaining the natural world. Universal gravitation worked on bodies, Newton had said, and its invisible force ultimately came from God. But why not have gravitation work *in* bodies, be a force inherent in bodies? Does the world not appear to move on its own devices? Does Newton's law not prove just that? Those were the questions that deists and atheists were raising at the moment when Clarke rose to attack them. Thanks to their reading of Newton, the deists had found a way of leaving the

Christian camp altogether. Indeed, in 1704 John Toland issued *Letters to Serena,* an impious work in which he mentioned Newton by name and said that the *Principia* proved that motion (or attraction) is inherent in matter.

This was no idle citation by a malcontent crank. Toland had been a student at the University of Edinburgh, where he had learned Newtonianism from David Gregory (1661–1708), one of its earliest disseminators. [30] Then, after more study in Leiden, Toland came home and moved into Whig political circles in London, where he was known all too well to John Locke. As he evolved toward materialism, Toland became a self-proclaimed "pantheist"; in fact, he even invented the word. [43] Toland was a man-about-town who addressed his philosophical writings to both men and women and traveled in high political circles. He had to be stopped.

Clarke moved quickly to attack Toland and to free his mentor's system from even the hint of association with atheism. In the pulpit of St. Martin's, under the auspices of the Boyle Lectureship, Clarke brought the weight of Newtonian science and the respectability of the Church of England down on the head of Toland and every other deist and atheist Clarke could cite. Newtonian science proved the efficacy of divine Providence, he asserted. "From the brightest star in the firmament of heaven, to the meanest pebble upon the face of the earth, there is no one piece of matter which does not afford such instances of admirable artifice and exact proportion and contrivance, as exceeds all the wit of man." [43:185–86]

Clarke's arguments were philosophically sophisticated, elegantly presented, and utterly dependent upon Newton's science. Their impact was so far-reaching that the rabbinical leader of London Jewry, David Nieto, who had just arrived in the country, took up arguments similar to those of Clarke and used Newtonian metaphors in his synagogue to show the hand of God at work in all of creation. [71] Religious thinkers affected by Newtonian metaphysics set the course of first British, and then Western, religiosity in a rationalist direction that separated the sensibility of the scientifically literate from the mass of the untutored populace. Physico-theology or natural religion became the fashionable alternative to traditional piety. In the circles of Italian intellectuals drawn to the new science, Clarke's Boyle Lectures became a new canon of orthodoxy to be used against the Italian versions of materialism, some inspired by Toland and his associates. Similarly, Clarke's epistolary polemic with Leibniz, once translated into French, was widely read and commented upon throughout the Republic of Letters. So famous were the disputants that the Roman Inquisition as the guardian of traditional piety added the French translation of the published Leibniz-Clarke correspondence to the *Index of Forbidden Books* while somehow managing to overlook Toland's far more dangerous exposition of Newton's concepts published in *Letters to Serena.* Sometimes censors have trouble distinguishing their friends from their enemies.

Aware of all these controversies, in the *Opticks* (1717) Newton put the relationship between matter and motion succinctly and as orthodoxly as Clarke and the other Newtonian theologians could ever have desired: "The *Vis inertiae* [the inertial force] is a passive principle by which bodies persist in their motion or rest, receive motion in proportion to the force impressing it, and resist as much as they are resisted. By this principle alone there never could have been any motion in the world."[18] In a scientific work on optics Newton did not go on to say directly that God put the motion into the universe, but the theologians led by Clarke said it, over and over again. Under his lead Newtonian natural philosophy gradually replaced Aristotleianism as a way of viewing the world that permitted science and could be seen to retain divine authority.

Newtonian Christianity, the marriage of Protestant theology to science, flourished on both sides of the Channel and in the American colonies until well into the nineteenth century. Clarke's Boyle Lectures were translated into many languages, and for believers his religiosity became a symbol of an enlightened version of Christianity that permitted science but turned its back on medieval theology and the sacrosanct clerical authority that went with it. The message of Newtonian Christianity focused on order in the heavens, stability and harmony on earth: "The sun's forsaking that equal course . . . would be to the natural world . . . the very same thing, injustice and tyranny, iniquity and all wickedness, is to the moral and rational part of the creation." [43:190] Clarke's religiosity became so commonplace that it would be unfair to trace the somewhat smug and self-satisfied understanding of the world common to eighteenth-century elites solely back to him.

For atheists Clarke's fame made him a target throughout the century. In the 1770s the leading French atheist and materialist, the Baron d'Holbach, attacked Clarke and simultaneously translated Toland's books and ideas into French. But when believers in God's existence who wanted his beneficence without recourse to the preachings of the entrenched clergy looked for arguments, they looked to Clarke and his followers. Even the famous French philosophe Jean-Jacques Rousseau, writing in *Emile*, expressed his sense of God's presence in nature by enlisting Clarke as his primary defense. But always the traditional and godly were suspicious of the Newtonian approach to Christianity. In the Dutch

Republic s'Gravesande watched over his shoulder to guard against attacks by orthodox Calvinists. At his inaugural lecture for his professorship in Leiden he attacked the detractors of science who spied in it the road to atheism and irreligion. His colleague Hermann Boerhaave also strictly maintained a rigid separation between mind and body, arguing that the attributes of the mind and body had "nothing in common with each other." [88] Although he was a practicing doctor, Boerhaave needed to be careful about the way he expressed what he saw so as not to imply that souls and bodies were one and the same thing.

The greatest fear among the eighteenth-century supporters of science focused on materialism. It conjured up the denial of free will and hence freedom; it also conjured up the equality of all matter and things—conceivably, when applied to human beings, a radically democratic vision of society. Churches and pious scientists were supposed to prevent such deteriorations in both public and private respect for order and for the Deity and his clergy. But even in scientific circles the temptation toward irreligion proved too inviting, at least so rumor in the 1730s had it. The horrified William Stukeley revealed that deep in the private dining rooms of the Royal Society its new president, Martin Folkes, and his "junto of sycophants" scoffed at religion "so that when any mention is made of Moses, of the deluge, of religion, scriptures . . . it generally is received with a loud laugh." [19:143] At least so the rumor went. Newton, Clarke, even Bentley would have recoiled. Yet for freer spirits science eliminated piety, and Newtonian physics made the universe look secure.

Newtonian religiosity seemed an excellent alternative to either traditional Christianity, with its fear of mechanisms and its support for witches and spirits, or to materialism, which implied a godless, predetermined universe. The price paid for Newtonian approaches to Christianity entailed certain key theological doctrines. Newton had been an antitrinitarian; so too was Clarke. As we saw in Part 1 (p. 53), Newton's religiosity gave a quasi-divinity to Christ, but the subtlety and purpose of his deep faith—to reveal the immediacy of the Creator—was alien to many of his subsequent followers. They sought instead to reveal the Deity by rationalizing His activity, by bringing the divine into focus because seen through rational and scientifically trained eyes.

By the 1730s so commonplace had Clarke's version of religion become that the French ambassador to London even reported on it in a secret report to his king. This was a rather optimistic document written about an old enemy by a diplomat who would have liked nothing better than to see the Protestant English dissolve into fractious sectarian warfare. The subject of the report was signs of dissension to be seen in the kingdom of France's main rival. The mention of Clarke should surprise us; his attempt, after all, had been to support the state and provide justification for harmony and stability. The French ambassador's report referred, however, to the new "sect" of the Unitarians and their debt to the theology of Samuel Clarke.

Unitarians (then and now) do not accept the divinity of Christ. Clarke's theology, with its emphasis on divine power and the indivisibility of the material atoms, and with its denunciation of "forms," had the impact, intended or not, of further undercutting belief in Christ's divinity. To be sure, neither Clarke, nor Newton, nor the Unitarians wished to foster irreligion. Theirs was rather a new, more cerebral religiosity that rendered all intermediaries between God and man less important. By the 1780s Unitarianism was widespread in both England and the American colonies. Thomas Jefferson even wrote a Unitarian version of the Bible. Jesters such as Erasmus Darwin, the impious grandfather of Charles, described this minimalist creed as "a featherbed to catch a falling Christian."

Throughout the eighteenth century high-church Anglicans spied in Newton and his followers the root of renewed heterodoxy that they believed threatened traditional authority in church and state. [48] They imagined a slippery slope that began with Newton's natural philosophy, leading its adherents to tumble in the direction of antitrinitarianism, winding up caught on the ledge of deism, or plunging into the abyss of atheism. For the orthodox, Newtonian science presented a no-win alternative: materialism or heterodoxy. For its many followers, however, it offered the magic formula that endorsed science and permitted a general, if vague, Christian piety.

High-churchmen found cold comfort in those few Newtonians, like William Whiston (1667–1752), who retained the intense and millenarian piety of Newton himself. Whiston took up antitrinitarianism because he believed it to be a purer form of Christianity; not least he knew that Newton, his great idol, had also denied the traditional definition of the divinity of Christ. Whiston also saw the end of the world approaching, and he went about telling people of it. For his candor about the Trinity Whiston lost his professorship in Cambridge. He too then began public lecturing. And his lectures never lacked drama; in 1714 he predicted that soon the world would end by being struck by a comet. Again the high-churchmen were horrified; "Whistonianism" joined the ranks of heresies abhorrent to orthodoxy. [63] To this day historian adherents of high-church orthodoxy have argued that true piety in eighteenth-century Britain lay elsewhere, in trinitarianism and a deep commitment to the traditional doctrines and prescientific arguments found among the rank-and-file clergymen of the established-by-law Church of England. [12] But in reality intellectual innovation and leadership within British elite culture had passed to the Newtonians and their followers, among whom must be counted the Unitarians. When Halley's Comet did return in 1759 it was seen as rationally explicable, as proof positive of Newton's system of the world. [73]

Eighteenth-century Unitarianism owed much to the theology of Samuel Clarke, and it became one of the most powerful of the new rational religions of the age. Perhaps the most famous Unitarian of the century, Thomas Jefferson may have first learned ideas similar to those of the Unitarians from his reading of another transmitter of Newtonianism, Henry St. John, Viscount Bolingbroke. In contrast to materialists, some of whom argued that Jesus had been an impostor, Jefferson embraced a philosophy that was "meant to place the character of Jesus in its true and high light, as no impostor Himself, but a great Reformer of the Hebrew code of religions." [11:27] To convey this Unitarian understanding of the man Jesus, Jefferson even rewrote the New Testament, expunging the parts that implied the divinity of Christ and dwelling on those parts that made Jesus a moral teacher. [11]

What Newton, the millenarian, would have made of Jefferson's adulteration of the Scriptures we can only try to imagine. There is no small irony in the adoration that Clarke at the beginning of the century, and Jefferson at the end, gave to the deeply pious Newton and his natural philosophy. In the hands of these interpreters, Newtonianism became the major intellectual force in the century's movement for reform and secularization known as the Enlightenment. Indeed, Jefferson actually belonged to the Enlightenment's leadership and can therefore be described as a major philosophe. By contrast, both Clarke and Newton rejected the secularism of the Enlightenment, already represented in their lifetime by the writings of minor philosophes like Toland. Yet in their way Clarke and Newton contributed to the formation of a new religiosity, what the age called natural or rational religion, but which we have just as readily described as a new version of Protestantism, as Newtonian Christianity.

The Variation of Freemasonry: New Rituals for a New Universe

At the heart of the Newtonian version of Christianity stood a deity described by Newton as the Great Mathematician; other Newtonians called God simply the Grand Architect. Inevitably the harmony of the

Newtonian universe inspired actual worshippers of the Architect, and these were found most readily in the new Masonic societies of the early eighteenth century. Imagining themselves as the descendants of medieval architects and stonemasons, they embraced the Grand Architect of the Universe, a Deity that some careful readers believed they could discern in the Scholia attached to the *Principia* and in the Queries published at the end of the *Opticks* (1706). Whole Masonic rituals centered around the Grand Architect, God of the new science, as well as around a Deity sufficiently vague that Catholics as well as materialists could be found in the new Masonic lodges that began in London and spread to every urban center in western Europe by the 1740s. [42] Freemasonry permitted religious toleration, support for science, and, not least, exercises in self-government and a new kind of secular sociability.

Britain was a country deeply divided by religious differences: non-Anglican Protestants did not have full legal rights; Catholics and antitrinitarians were legally excluded from toleration, although largely left unbothered in practice; and there were dozens of small sects, Quakers, Baptists, and Congregationalists, as well as French Huguenot refugees. To find a way through this thicket of religious tension, tolerant-minded Newtonians like Desaguliers and Stukeley, along with a few Whig gentlemen, invented a new form of ritual to worship the Grand Architect of the Universe. In offering an alternative to established religion they also sought to leave men to their individual doctrinal and sectarian beliefs. Out of the old guild fraternizing of working stonemasons, who now left their guilds in the face of a free market in wages, their gentleman replacements fashioned a private fraternity, the new Freemasonry. In the new lodges gentleman freemasons worshipped the Grand Architect while practicing the institutions of constitutional government, holding elections, giving speeches, and ruling by majority votes.

By the middle of the century a working mason would barely have been welcome in most lodges; this was a sociability for the literate, even for the elite. In town after town the lodges also assisted the middling classes to see themselves as separate from "crude" working men and sufficiently genteel to break bread with their aristocratic and gentry betters. The educated and literate wanted to know the traditions of artisans. They also wanted to employ them at wages set by profit-seeking entrepreneurs and engineers.

Of the many innovations that sprang from Newtonian inspiration, Freemasonry seems fanciful, even arcane, and yet it was by far one of the most popular forms of sociability (with religious overtones) found in an age devoted to pubbing and clubbing. By 1750 there may have been as many as 50,000 freemasons in Europe. It was as if the order and harmony inherent in the Newtonian model of the universe had inspired a new model for society, one that also consciously identified with the new constitutional English style of government established by the Revolution of 1688–89. In the lodges men (and on the Continent eventually some women) could find a society governed by laws and constitutions while worshipping a model of order and harmony inspired by Newtonian science.

The original masonic lodges had been simply local meeting places for stonemasons, some of whom were mathematically skilled as well as being practical architects. There is even some evidence that the great architect of the Restoration, Christopher Wren, had been made one of the first Grand Masters of the London lodges. But most practicing masons were far less knowledgeable and skilled than the architect of St. Paul's Cathedral. The seventeenth-century guilds in Scotland and England have, however, left evidence that early stonemasons possessed intellectual interests, and these included the belief that the original architect-masons had worked on the Temple of Solomon.

The Temple was an edifice to which Newton also devoted his scholarly time. He believed that in its proportions might lie one clue to the cosmic calculations of the Deity. Some of his calculations were even published after his death in chapter 5 of his *Chronology of Ancient Kingdoms Amended* (1728). All these associations with Newtonian religiosity and science inspired Masonic admiration and imitation. The building of the Temple, and mechanical knowledge in general, were seen as evidence by men like Desaguliers and Stukeley that even in the Middle Ages those who knew mathematics, especially geometry, were closer to understanding the very nature of the universe.

That Newton was impressed by the mysterious proportions of the Temple of Solomon could only have aided Masonic recruitment. In the lodge a brother could come closer to the true primitive wisdom that lay at the root of all religions and at the source of cosmic order. Little wonder that one of the most devout Freemasons to be found in the early lodges, William Stukeley, was an ardent supporter of Newtonian science. He spent years of his life trying to reconstruct the mathematical and mystical proportions of the original Temple while faithfully attending meetings of the Royal Society, recording his admiration for Newton, its president, and also being a devoted lodge member. In his love of science, antiquities, and history, Stukeley belonged to the rather tame version of the Enlightenment that became commonplace in Great Britain. To this day his voluminous writings on the Temple of Solomon sit unpublished, donated to the Freemasons' Hall in central London. To modern readers they seem eccentric; to his fellow lodge brothers and Newtonians they were part of the search for an original, mysterious not so far recovered wisdom. Often on the same piece of paper he would record things that Newton said at a Royal Society meeting and an imagined proportion for the Temple of Solomon.

Sometimes rationalism takes forms unrecognizable from one age to another. Even while dabbling in ancient temples and myths, most Masonic lodges self-consciously sought to be ideal and ordered societies where discipline, harmony, and the rule of law prevailed. This idealism also picked up on the old guild ideals of social equality and on republican ideology as found in the first English revolution. It proclaimed that merit should be the sole criterion for determining social place and esteem. All these ideals—equality, merit, fraternity, order, harmony, and the perfectability of individuals and society—were woven into the beliefs of Freemasonry.

The rapid European spread of the Masonic lodges gives dramatic testimony to the eighteenth-century search for a way out of traditional culture, centered as it was around church, guild, or social estate. In the lodges the Enlightenment was lived by ordinary men and some women who read science, bought books, went to lectures, and imagined that society could be as ordered as the Newtonian universe. So potent was the Masonic message, and so nontraditional were its ideals, that in the 1790s monarchists and the supporters of absolutism in church and state blamed the French Masonic lodges for having caused the French Revolution. The charge was nonsense. But the European lodges did offer some men, and the few women who joined them, a new and satisfying cultural experience, one that they believed possessed universal validity as well as scientific justification. The harmony of Newton's magnificent universe inspired imitation in society and government. At the very least its proven order and stability could be worshipped by literate elites who, regardless of how little science they actually knew and if only for an evening, turned themselves into priests officiating at ceremonies invented to honor the Grand Architect of the Universe and his magnificent creation.

European Scientific Innovation under Newtonian Inspiration

By the 1740s much scientific activity throughout western Europe had come to be dominated by experiments and mathematical exercises derived from the *Principia*. Its elegance and experimental aesthetic gave some men and women religious inspiration; it also informed the life work of various mathematicians and experimenters. But what became highly original and innovative work in Newtonian science did not occur in its country of origin. In the eighteenth century British science took an increasingly applied and industrial direction, which, as we shall see in the final section, had implications for the whole of British and eventually Western history.

Throughout much of the eighteenth century the gentlemanly membership of the Royal Society of London coalesced around the utilitarian as well as the antiquarian. [56] Even in Newton's lifetime the Society had turned toward the applied—some might have said the pedestrian—in matters scientific. Local artisans would even come there to display their work. Clocks, pumps, castings of water pumps, models of engines, musical instruments threatened to overwhelm even electrical experimentation. The majority of the Royal Society had always been genteel and curious, with highly skilled artisans like Benjamin Martin coveting membership but frequently being rebuffed. By midcentury collecting, surveying, and applied mechanics—all with implications for industrial development—became the order of the day in British science. [10] Finding conceptual and mathematical innovation, the stuff of traditional histories of science, meant crossing the English Channel.

When in the 1780s Thomas Jefferson fulfilled his dream of going to Paris, he wrote back to his nephew that he should practice French to perfection "because the books which will be put in your hands when you advance into Mathematics, Natural philosophy, Natural history, etc., will be mostly French, these sciences being better treated by the French than the English writers." [87:681] Knowing what we now do about English science in Newton's lifetime, this seems an extraordinary statement. Jefferson was, of course, partial to the French philosophes of the Enlightenment, Montesquieu and Rousseau in particular. But he was telling his nephew about science, and he was basically right.

Innovation inspired by the *Principia* appeared in French science as early as the writings of the priest and mathematician Pierre Varignon. Between 1695 and 1715 he used Leibniz's calculus to express the motion of moving bodies in a resisting medium—work directly inspired by Book II of the *Principia*. Varignon launched mathematical dynamics and mechanics in France, but the next generation brought them to new and extraordinary levels of originality. Among the most original and important thinkers of the second half of the eighteenth century towers the French scientist Pierre-Louis Moreau de Maupertuis. In 1738 he became the first person to advocate the Newtonian theory before the prestigious French Academy of Sciences, and he was the first Continental scientist to use Newton's theory of gravitation to establish the shape of the earth. His work is exactly contemporaneous with the more practical engineering and lecturing of Desaguliers.

Maupertuis's academic appearance was preceded by decades of contact between British Newtonians and French philosophers, mathematicians, and journalists. Indeed, Newton was known by reputation in France from the 1690s. But actual Newtonians, capable of challenging the dominant Cartesianism of the colleges and academies, appeared only gradually. First came interest in Newton's theory of light and colors. Yet his famous prism experiment was difficult to replicate successfully, and French scientists argued that Newton's optics were just plain wrongheaded. They simply respected him as a mathematician and an experimenter. There were Cartesian scientists and mathematicians who did work with the calculus. Relations between bodies could be calculated more easily, and that enhanced understanding of motions and magnitudes. [29] Cartesian mathematicians even submitted papers to the French Academy demonstrating that a hyperbolic spiral path implied a central force proportional to the cube of the distance. By 1700 French scientists were moving toward the mathematics of universal gravitation, just as earlier they flirted with Newton's optics. Yet they wanted to avoid the basic truth that in matters celestial you could not be both a Cartesian and a Newtonian. Eventually the vortices were seen as nonsense by Newtonians and, as we saw in Part 1, they are central to Cartesianism. With their demise the Cartesian house, although not the Cartesian spirit of rationalism and deduction, crumbled into ruins.

Assisting its demise were various Newtonian fellow travelers. The mathematician, musician, and freemason Brook Taylor (1685–1731), along with the exile and sudden convert to Jacobitism Henry, Lord Bolingbroke, went to the land of the Cartesians and brought Newton's science with them. Bolingbroke was a deist and cared little about the religious implications of Newtonian natural philosophy, but the pious Taylor claimed that Cartesianism threatened sound, natural religion. Taylor was also a practicing Newtonian, and his collaboration with Hauksbee had led to an effort to determine the laws of magnetic force and capillarity (the movement of liquids in tubes and pressed surfaces). As an ally of the great but ruthless Newton, Taylor played the nationalist game and agreed with the Royal Society when it claimed that Newton, not Leibniz, had invented the calculus. There was national interest but also Whiggery at work in the defense and promulgation of Newtonianism. [75] It just would not do to have the new Hanoverian king come over with the scientific archrival in his German retinue. How better to secure constitutional monarchy and oligarchic rule than to make the very heavens the source of law and order and to ensure that Newton be seen as the very font of all mathematical wisdom. Hardly surprising that there were French Cartesians who regarded Newtonian science as "a national malady that no remedy can cure; thank God it has not crossed the sea."

But cross it did, and in the 1730s the tide turned irrevocably in a Newtonian direction. Voltaire helped with his *Letters,* but Maupertuis was the key figure. First his mathematical competence commended him to other mathematicians, then he achieved international fame by a trip taken to Lapland to measure the shape of the earth. Seen as a crucial test of Newton's theory of gravitation, the experiment done by Maupertuis successfully proved that Newton had been right when he argued in the *Principia* that the rotation of the earth coupled with the force exerted upon it by the gravitational pull of the sun should cause it to bulge at the equator and be flattened at the poles. In Newton's lifetime some evidence for the proposition existed: it had been discovered that a pendulum swung more slowly at the equator than it did in France. If the earth is thicker there, the device would indeed be further from the source of gravitational pull, which lies at the center of any body; hence it would be pulled more slowly toward the earth. Maupertuis and a band of scientific followers, among them a clockmaker, undertook a perilous polar expedition, got as far as Lapland, and there established the significant differences between pendula swung at different points on the earth's surface. The earth turned out to be more like an onion than a football, and Newton turned out to be right.

Maupertuis's fame brought him to the attention of the Prussian king, Frederick II, who put him at the head of his new Berlin academy. Indeed, Frederick specialized in acquiring Newtonian mechanists; Voltaire was another one of his temporary visitors. So too was La Mettrie, the materialist. It was all part of bringing the Enlightenment to Berlin in order to enhance monarchical power. The project of advancing Newtonian science had now become an international competition, and the Prussians, along with the French, became key players in the field. By the 1750s the Russian Academy of Sciences also entered the fray by offering a prize for the best paper on the apogee of the moon, the exact calculation of the moon's orbit at its furthest from the earth: no mean feat, because the difficulty lies in calculating the mathematics of movement for a body being pulled strongly and at different angles by both the earth and the sun. Newton had said that the problem literally gave him a headache.

Now, decades after Newton's death, a French mathematician and rival of Maupertuis, Alexis Claude Clairaut (1713–65), laid claim to the exact mathematical solution to the apogee problem. This midcentury generation of French Newtonian mathematicians pulled off another dramatic coup when Clairaut also predicted the return of Halley's Comet within a month of its actual appearance in 1759. The French press said that Clairaut was the new Newton, while another Newtonian mathematician and leader of the French Enlightenment, Jean d'Alembert (1717–83), sulked in envy. He thought himself to be the scientific leader of his generation.

D'Alembert was a leader in his generation, although his fame came to rest most dramatically not on any equation but on a single, long preface to the first major encyclopedia in Western history. In collaboration with Diderot and a consortium of writers and publishers, d'Alembert sought to apply the method and goal of the new science to classify and codify all knowledge and bring it together in encyclopedic form. The now world-famous result was the *Encyclopédie* of 1751, the forerunner of the modern genre. It explained Newtonian science along with most other important innovations of the age, not to mention heresies—all so beloved by the leaders of the Enlightenment. The new encyclopedia became their most essential and representative text. The article on "attraction" explained that "one of the accepted rules of philosophy [is that] the weight of any body is proportional to the quantity of matter in each body . . . and it is a law of nature that each particle of matter tends toward every other." A natural universe bound by mathematically knowable laws, capable of human imitation and composed of universal truths, became not only possible but real. The universalism of enlightened ideals—religious toleration, freedom of thought, human rights in general—cannot be separated from the universality of Newtonian mechanics.

The generation of French mathematicians led by Maupertuis and d'Alembert brought Newtonian celestial mechanics to new levels of sophistication, and they elaborated upon the *Principia* in ways that further consolidated the new mathematical physics. They should not be imagined as working in isolation or removed from a very specifically Continental, politically absolutist context. The state-sponsored Continental academies of science, with their aristocratic complexion, fostered theoretical and mathematically original work. [51] They did so far more readily than their British counterparts, filled as they were with amateurs and gentlemen. The social settings of Continental and British Newtonian mechanics and physics helped to move them in different directions. The first moved toward the theoretical; the second, toward the practical and the applied. [44]

There were other noticeable differences between the directions in which Newtonian science was pushed by French and British commentators, respectively. Georges Louis Leclerc, Comte de Buffon, embraced

Newtonian physics but would have nothing to do with the theological underpinning of universal gravitation. The natural religion or physicotheology so beloved by British Newtonians was far too godly for Buffon and the leaders of the French Enlightenment. God may have set the planets in motion, but this was now irrelevant. Buffon's *Theory of the Earth* (1749) and his subsequent work used mathematical probabilities to express the regularity of planetary motion. All the planes of the planetary orbits are inclined no more than 7 $^1/_2$ degrees from the ecliptic, and the probability is 7,692,624 to 1 that this inclination could not have been produced by accident. But while Newton looked to God for explanation of such regularity, Buffon looked to the heavens for a solution. A single comet, he concluded, by falling obliquely on the sun, dislodged a piece of it. This matter would be in a liquid state and in a torrent would fly vast distances. The smaller and more dense particles would remain nearer the sun, and the power of attraction built into all matter would operate on all these detached parts. Other French followers of Newton accused Buffon of plunging back into the obscurity of hypotheses. But late in the twentieth century we can recognize in Buffon's efforts the first attempt to provide a purely scientific explanation for the origin of the solar system. When Buffon's comet hit the sun, he might now say, it too had produced a "big bang."

Amid these speculative innovations in Newtonian mechanics, theoretical mathematics and mechanics remained vibrant among French Newtonians up to, and beyond, the French Revolution. Pierre Simon Laplace (1749–1827) rejected Buffon's comets and argued instead that the planets had been created from the atmosphere around the sun, which had extended vast distances. As this atmosphere condensed, a succession of rings remained, still within the plane of the sun's equator, and these in turn coalesced to form the various planets. In a similar fashion the planetary atmospheres in turn produced satellites or moons. The cohesion needed to produce the planets was by no means a certainty; some bodies would remain relatively unformed asteroids. [60] When William Herschel in England used the new and most powerful telescope then available to discover a new star surrounded by a "faintly luminous atmosphere of a considerable extent," Laplace said that at last the evidence had been discovered to support his purely scientific, naturalistic explanation of the universe. Newton would have been horrified; even Napoleon thought that Laplace ought to say at least something about God. William Paley (p. 187) denounced the whole French effort. In more than 40 American editions of this British theologian's famous textbook in natural philosophy, students learned that Laplace and his followers engaged in "the grossest philosophical absurdities."

But by 1800 the entire direction of Newtonian physics and mechanics, both theoretical and applied, had become vastly more sophisticated; critics would have also said "godless." Two centuries earlier the great controversy had been about whether or not the earth even moved. In 1633 Galileo had been condemned and put under house arrest for being a Copernican, for accepting the truth and elegance of the heliocentric model of the universe. In 1757 the church still opposed Copernicanism but did manage to avoid prosecuting its proponents by abolishing the anti-Copernican decree of 1616. This was a belated locking of the barn door, because by 1757 Newton's work was widely known in Italian scientific and intellectual circles.

By 1800, under the impact of the *Principia*, mechanics, now internationally practiced and communicated, had become so refined that experimenters used microscopes to measure the exact deviation due to air resistance experienced by falling bodies. Perhaps in the final irony of the century, mechanical experimenters of the 1790s sought permission to drop their weights from the splendidly high tower of St. Peter's Basilica in Rome. They would do this at night so as not to disturb the pilgrims. The local cardinal who supported the project was otherwise too politically embroiled to bring it to fruition, and a tower in Bologna worked out just as well. [53] Whatever the outcome, had the owl of Minerva, the companion of the goddess of wisdom, known about the request, the coy wink for which the owl is famous could only have been accompanied by an even slyer grin. The universe had been mechanized, even in Rome, the very bastion of the kingdom that good English Protestants like Newton and Boyle would once have identified as belonging to the anti-Christ.

The British Practice of Newtonianism: The Cultural Foundations of the First Industrial Revolution

Far from the purely mathematical genius of a Maupertuis, or from the ceremonies and rhetoric of the Masonic lodges and the worship they gave to the Grand Architect of the Universe, lay the everyday world of machines, industry, and work. In that setting and late in the eighteenth century, Newtonianism as a system for organizing applied mechanics made its greatest impact. Put succinctly, British applied mechanics, organized under the rubric of Newtonian science, was significantly more advanced and had penetrated society more deeply and widely than in any Continental country. [41] That startling cultural advantage became one of the main noneconomic causes of the First

Industrial Revolution. When we ask why Britain industrialized first, one of the answers lies in the cultural legacy of Newtonian science and its applications. [57]

Using Newtonian science taken from those parts of the *Principia* pertaining to the mechanics of local motion, Newtonian commentators, engineers, and teachers created curricula and books applicable to technological innovation. In the Royal Society of London, but especially in numerous provincial scientific and philosophical societies, this mechanical learning formed the centerpiece of discussions, demonstrations, and lectures. [27] Into this setting of not only formal but, just as important, informal institutions for applied scientific learning came eighteenth-century entrepreneurs, engineers, governmental agents, even skilled artisans, faced with economic and technological choices and receptive to new knowledge systems that promised new solutions. [62] The route out of the *Principia* (1687) to the coal mines of Derbyshire or the canals of the Midlands was mapped by Newtonian explicators who made the application of mechanics as natural as the very harmony and order of Newton's grand mathematical system. They operated within the forum of civil society, a vast interlocking network of private voluntary associations, public lectures, and informal study groups to be found in almost every provincial city and town. [17] The mathematical and mechanical practitioners often combined their skills with traditional values and attitudes, but most important was the certainty that the *Principia* offered learning about the heavens and about everyday motions accessible to anyone who could master a handbook.[19]

Recent historical research has re-created the setting in which mechanists and entrepreneurs first made possible the application of Newtonian mechanics, hydrostatics, and pneumatics. They did so frequently in partnership, the entrepreneurs consulting with civil engineers with whom they could compete in the area of general mechanical knowledge. In a space filled by capital, mechanical know-how, cheap labor, technological trial and error, and the prospect of domestic consumption, decisions were made that, taken slowly and cumulatively, resulted in the industrialization of mining, transportation, and manufacturing. Laid to rest forever is the older image of the industrial entrepreneur as a semiliterate tinkerer. For much of this century this benighted fellow permeated the literature on modern industrial development as well as the economic historiography of Western industrialization. In his place the cultural historian now points to the relationship between civil engineers and entrepreneurs, joined together by a common scientific vocabulary, allied by commonly perceived interests. Also laid to rest is the notion that "the great discoveries of mathematical physicists were not merely over the heads of practical engineers and craftsmen; they were useless to them." [31:334] That judgment could not be made after a close reading of Desaguliers or s'Gravesande.

After the publication of the *Principia,* but at varying rates of availability, applied mechanics became an organized branch of science, presented in textbooks found by the mid-eighteenth century in all the major European languages. But it was not just the books, it was the culture of science that contributed decisively to industrialization. The public culture of British science was markedly different from that found in other parts of western Europe. A scientifically inclined young man could learn the basic principles of Newtonianism from parts of the *Principia* or the *Opticks,* or from books about them taught by lecturers such as Desaguliers. He could also join a local philosophical society, or set up his own study group as did the young chemist Josiah Wedgwood, later to be famous as the maker of fine china. [39] Or he could learn mechanics at a grammar school (the equivalent of a high school) decades before that knowledge was routinely offered in the roughly equivalent French college.

The young John Grundy, who drained the fens and improved the rivers of eastern England, said that he had been inspired by hearing Desaguliers at a meeting of the Spalding Society. [65] Such a would-be imitator of the Newtonians might aspire to become a professor of physics but lack the means of attending university, even of affording membership in the Royal Society. Yet John Smeaton, who was such an impecunious youth, managed to find work on canals, harbors, and coal mines and in effect to invent for himself the profession of civil engineer. First he was trained in Newtonian mechanics by a relative, then he became a self-employed mechanical entrepreneur, a businessman-engineer who regarded himself as a physicist and scientist. [41] He was in fact simply a good Newtonian practitioner lucky enough to exist in a social setting where publication was uncensored, public lectures and books were easy to come by, professional life was not monopolized by the engineering schools run by the government, and decades of prosperity had already shown the value of coal deposits and good internal transportation routes. Not least, the provincial scientific societies allowed their members like Grundy to imagine themselves as citizens of a worldwide scientific effort. When in the 1720s the Royal Society organized an international weather survey that enlisted Italian Newtonians, Swedish theologians, and Harvard professors to contribute data, there were the Spalding gentlemen in deepest Lincolnshire eagerly contributing data from their none too precise thermometer. [34]

Indeed, all the available evidence now suggests that throughout the eighteenth century there were significant differences between the British setting and every other European setting where scientific knowledge of an applied sort can be found. These cultural differences were both quantitative and qualitative in nature, but only a comparative perspective throws them into sharp relief. Being comparative about culture allows for the distinctive or indigenous in each setting to emerge. It also tackles from a cultural perspective one of the most perplexing aspects of early industrial development, namely, France's relative technological backwardness. [78] By briefly examining Dutch and French cases in relation to the British, the historian can also add political elements, for each country possessed markedly different political systems, from absolutist bureaucracy to decentralized, locally controlled republic. [38]

Eighteenth-century Britain possessed a distinct scientific culture based upon applied Newtonian mechanics and technological application. This culture inculcated experimental methods, the discipline and style of replication and verification, and brought these practices to technological problems. Its participants simply could not have understood the sharp distinction made in modern times between the scientific and the technological. A letter of 1778 to James Watt about the steam engine he had invented, from the then famous civil engineer John Smeaton, makes the point nicely: "Resolving, if possible, to make myself master of the subject, I immediately resolved to build a small engine at home, that I could easily convert it to different shapes for Experiments. . . . I determined to prosecute my original intentions of finding out the true *Rationale.* . . . The fact is . . . I have no account upon which I can depend, of the actual performance upon a fair and well attested experiment, of anyone of your engines. . . . If you can shew me a clear experiment to this amount . . . I should think it no trouble to go to Soho [Watt's workshop] on purpose to see it."[20]

With these disciplined methods of verification and replication drawn from scientific practices, British engineers believed themselves to be scientists, or at the least their imitators. They could move effortlessly from hands-on knowledge of machines to the application of theories drawn from mechanics, hydrostatics, or pneumatics. [69] Not least, science and mathematics occupied their leisure and informed the education of their children, and they bought books and instruments in all fields from optics to astronomy and telescopes. [83] Their mental posture could be described as a median between theoretical science and highly skilled artisanal craft. They knew machines from having built them or having closely examined them, and they could relate their working to basic theories in mechanics, hydrostatics, and dynamics. They could use the same principles to drain mines, to calculate the size of an engine needed at a manufacturing site, to lay canals over hilly terrain, or to restructure harbors. They could turn to handbooks by James Ferguson, for example, that offered in clear and simple prose basic descriptions of steam engines, complete with diagrams.

The scientific knowledge engineers and entrepreneurs possessed was not, however, exclusively theirs. It came from the courses given by traveling lecturers, from patient study of textbooks based upon the *Principia,* or from regular attendance at the proceedings of the local literary-philosophical society such as the Lunar in Birmingham. There members could begin their studies by calculating the distance needed to offset disparate weights placed on a beam, go through levers, weights, pulleys, and engines, and end with a verbal and pictorial description of the Newtonian universe as explicated by the law of universal gravitation. In one exceptional town of 500 families, Spalding in Lincolnshire, local enthusiasts of science set up a philosophical society composed of medical men, clergy, local gentry, lawyers, a local painter, musicians, an ex-slave, some tradesmen, and women visitors. Led by a Fellow of the Royal Society, easily a dozen lecturers, among them Desaguliers, taught applied mechanics, gave electrical demonstrations, and made science as fashionable as sermon attending. [85] Among the members of this society were leading Newtonians of the first generation, men such as William Stukeley who had known Newton personally and listened to him speak at the Royal Society. These same men demonstrated an active interest in industrial machinery; they visited steam engines and reported on the principles of their operation. [69]

Thanks to the lecturers at work in towns, cities, and philosophical societies, by the mid-eighteenth century the British had the edge, relative to other countries, in knowledge of basic and applied mechanics. "Dr. Desaguliers who was my master, was the best engineer we ever had and has left ye Best Instructions of any I ever read, though there is as good French authors": so wrote an English engineer in search of a contract to drain the marshes around the French port of Dunkirk with his Newcomen engine. He was eager that his English experience and the reputation of his Huguenot refugee mentor would precede him. Little could he have known—as we now do—that the French ministries were deeply concerned about English technological prowess. In their private reports the ministers even said that it was due in part to the large Huguenot population that had migrated to England.

The exceptional penetration of science in Britain must now be reckoned as one of the necessary conditions permitting the first turn to industrialization. The "mental capital" of the First Industrial Revolution belongs with a cluster of key economic factors that worked to the advantage of entrepreneurs and engineers. Within the resources provided by the new scientific culture they could acquire applied mechanical knowledge made explicitly and directly applicable to technological decision making. Put concretely, it was possible to learn more about applied mechanics at a London coffeehouse lecture series than it was in any French college prior to the 1740s. [9]

In the degree of penetration and application of Newtonian science the British were easily a full generation, if not more, ahead of all other Europeans or Euro-Americans. Only in the 1740s did the curriculum of the nearly 400 French colleges shift decisively away from Cartesian metaphysics toward both a theoretical and applied Newtonianism. Focusing on the most backward of the 400 colleges in eighteenth-century France, L. Brockliss concludes that "if Newton finally triumphed in France it was probably over the corpse of the Jesuit Order." It was only during and after the democratic revolutions late in the eighteenth century that the curricula of western Continental schools and universities shifted decisively in the direction of application, giving less emphasis to purely theoretical mathematics and science.

Yet by the third quarter of the eighteenth century Newtonian mechanics were widely taught in the elite French engineering schools intended for military men. In Russia a copy of the *Principia* (1687) turned up only in 1718, brought by the chief physician to Peter the Great, and it was decades more before Newtonianism penetrated the schools of military engineering and the academies. [47] In the Dutch Republic Newtonians such as s'Gravesande and Peter van Musschenbroek believed that Newtonianism would flourish "even more but for the resistance of certain prejudiced and casuistical theologians." [32:32] The Dutch academies did teach Newtonian science, but fitfully. The dissertations done in physics and natural philosophy, while Newtonian, were largely mathematical and seldom applied. By contrast, in the 1720s a young artisan like John Watt (related to the famous inventor of the next generation, James), then in his early 20s, possessed a good working knowledge of rudimentary mechanics.[21]

For the same period evidence from the highly literate Dutch Republic demonstrates a noticeable retardation in applied scientific education offered in Dutch colleges and a concomitant lack of interest in applied mechanics on the part of Dutch elites. [45] When in 1790 James Watt was negotiating to send a steam engine to the Republic, J. van Liender, the leading importer of engines into the Republic, advised him to "give as much explanation as possible and a great deal more even as you did to that of the Batavian Society's Engine because everyone there shall understand so little of the matter."[22] In one of the main Dutch philosophical societies of the period, when all the necessary technical knowledge lay in accessible published texts, there had not been sufficient interest in the new technology to warrant its mastery.

You cannot deploy that which you cannot understand, and widespread educational gaps limit choices. Accessing new knowledge systems requires receptivity, and published accounts of Watt's engine in Dutch, as well as working models of the engine, were available to be seen and studied at leisure. But all the evidence so far collected suggests that there was a noticeable gap in the depth and breadth of assimilation accorded that knowledge even in the urban, majority Protestant, relatively free, and highly commercial Dutch Republic as compared to the British situation. Its mercantile elite valued astronomy for navigation, but not applied mechanics for manufacturing industries. [15] By the mid-eighteenth century the libraries of some Dutch technical colleges were noticably deficient in applied mechanics. S'Gravesande's successor to the chair of physics in Leiden, J. Allamand, possessed many scientific interests but was not particularly specialized in any of them. His friends said that he preferred politics and physico-theology.[23] In addition, other evidence suggests that the oligarchic and Orangist elite that controlled most political offices and institutions in the Republic were indifferent to industrial development.

Amid the political turmoil of 1788 in the Republic, Van Liender wrote to Watt: "Were public circumstances in another turn, than they now are, the Steam Engine would undoubtedly take footing in this Country; but by being a work of Patriots it is quite condemned and abhorred." [45:238] In the Republic the turmoil did not end until 1815, but by then Dutch retardation was marked by comparison to the southern, Belgian provinces. The new united Kingdom of the Netherlands incorporated into the old Republic the existing areas of advanced mechanized production in Belgium and favored their further development. This policy meant that when in 1830 the Belgian revolution succeeded and severed those provinces from the former Republic, it further exacerbated industrial retardation in the northern Netherlands. [28]

None of these political or cultural details used to be of much interest to the mechanical models that still permeate historical writing about industrialization. Such economic models simply assume that if people have coal, capital, and cheap labor, they will see it as being in their best interests to industrialize. If they need any specialized scientific knowledge to do that, they will just go out and get it. [13] Such arguments about the way human beings change, make choices, or even recognize what choices are available presume a particular definition of the way people are. Their free will transcends any restraints that culture or circumstances may place on them. Rationality means always choosing what is perceived as being in one's best interests. Put another way, offer someone the chance to make a profit—in this case, to industrialize—and they will opt for progress, do anything, invent or innovate, try and try again until they succeed.

To clear a path through such simplistic arguments and to make way for a cultural and nonreductionist reading of the sources and circumstances of Western industrial change, the contemporary cultural historian must also grapple with the voluntarist ("just say yes") or instrumentalist ("see the end, just choose the means") frameworks of analysis inherited from an older historiographical tradition best exemplified in the writings of such major commentators as John U. Nef, Douglas C. North, and Robert P. Thomas, refined and elaborated upon by historians such as David Landes. The voluntarist approach and instrumentalist "rational choice" models of human behavior share the same set of conceptual problems. They convert culture into a set of rules, followed or transgressed depending upon perceived need or self-interest. As a result of the rigidity of this model, it cannot account for "failure," for why certain societies do or do not industrialize, or do so at different times. The model allows little explanatory role to policies, social systems, cultural values, scientific sophistication, levels of educational attainment, or forms of government. The so-called rational choice model of human actions would see little of interest in the differences between the various scientific cultures that emerged in eighteenth-century Europe. They would look elsewhere: for example, solely at supplies of capital or cheap labor to explain Britain's extraordinary leap forward in mining, transportation, and manufacturing.

In the rational choice model of industrial history, failure or success was simply the aggregate result of innumerable and discrete moments of choice made by entrepreneurs as they calculated their perceived interests. Or, in another version of this mechanistic model of historical change, progress or retardation becomes a matter of chance, the random motion of capital and goods. The role of culture—imagined as the tinted spectacles that enhance or impede individual perception, or that sharpen short-range or long-range vision—has no place in either model of human agency. But showing the marked differences between the scientific cultures found in Britain by comparison to France or the Netherlands tries to re-create the different universes wherein entrepreneurs actually lived, and from there to suggest that these cultural universes played a historical role that was important. Cultural arguments should never supplant economic ones, but then neither should the reverse kind of reductionist explanations be allowed to dominate the effort to understand the origins of modern industrial societies.

The tenacity of rational choice models of human conduct results from their long lineage in Western thought and, perhaps ironically, from their debt to the very scientific and mechanical culture here introduced into the explanatory framework of Western industrialization. In effect mechanical models assign to human actors the automatonlike qualities of Vaucanson's ducks. The rise of mechanical models to preeminence was closely related to the success of late seventeenth-century English and Newtonian science and its particular relationship to notions of freedom. [40] Vindicated by the triumph of parliamentary and constitutional systems of governance and relative religious toleration, the voluntaristic model of free agents freely choosing their economic destiny along with their religion and monarch became an ideological component within scientific culture itself. Newton's contemporary William Petty, the first political economist, who invented his discipline at that revolutionary moment, put the relationship succinctly: "Liberty of Conscience, registry of Conveyances, small customs, banks . . . rise all from the same spring."[24] Religious liberty came to be equated with economic freedom—for those with some wealth and skill.

With an interest in both the religious and philosophical aspects of the spring of freedom and wealth, the Royal Society of London as early as the 1680s discussed the labor-saving value of machines.[25] But for an inventor or entrepreneur to get a patent in Britain even in the 1740s, the bias of the authorities was overwhelming toward the argument that a device would put the poor to work, not enhance profits by reducing labor costs. [50] Indeed, Desaguliers's 1744 textbook in mechanics, *A Course* . . . , offered a discussion of the steam engine that was the first spelling out in print (vol. 2, p. 468) of the critical insight that mechanization undertaken by engineers could enhance profit precisely by reducing labor costs. To do that, he lectured, required good technical knowledge and sound engineering.

Desaguliers's explanation of the keys to industrial success picked up on what earlier seventeenth-century English theorists of political economy such as William Petty had explained. They equated free choice with the profit motive and with calculated striving to do what needed to be done in order to succeed. Freedom, choice, and success were the antithesis of idleness, and they even had nothing to do with the selling of one's labor. [7] Ownership and manipulation of goods or objects by employers represented choice, the natural activity of free men. The voluntarist model took literacy and technical learning for granted, its

beneficiaries having in many instances acquired both through a literature that extolled various Protestant virtues of work, discipline, and
preordained, God-given order. Science, discipline, ownership, and freedom were united in the creed that Petty, Boyle [40], and their Newtonian
successors tried to instill. With a confidence born in both politics and
science they believed that "the whole system of the World was made
for the use of our earth's men."[26] The behavior of the landed and
propertied came to be understood as the natural condition of all humankind. The arguments of scientists and theorists like Petty and
Desaguliers became so much a part of Western culture that amnesia
set in about how these arguments had been constructed and how distinctive and original they were in their country of origin.

The very strength, and eventual success, of the new scientific culture gave voluntarism or "rational choice" many inbuilt advantages, as
well as considerable staying power to this day among the inherited
Western intellectual paradigms. By its very success, the rationalist model
of human behavior obscured its origins within the scientific culture
inherited from the seventeenth-century English revolution in science
and government. When teaching about mechanics and experimentalism,
the Newtonian lecturers of the eighteenth century were doing something that was historically new—not in the very nature of humankind—
and they were reinforcing the projecting, entrepreneurial interests of
many of the men and women in their audience. They officiated at a
marriage of convenience between engineers and entrepreneurs: "The
contriver was a curious practical Mechanick, but no mathematician nor
philosopher; otherwise he would have been able to have calculated the
Power of the river." Had he or another properly trained engineer done
so, Desaguliers concluded in his second volume of *A Course* (pp. 530–31),
the correct management of power would directly have reduced costs.

What was being formulated in eighteenth-century Britain was an
industrial culture. At its social center lay the articulation of a new
relationship between the engineer and the craftsman. At best it had
been uneasy; for centuries both parties to it had conceived of their
skills as separate and unrelated. But beginning with Galileo and his
work in mechanics, a redefinition of the relationship was embraced by
theorists. They said: Learn from practical, uneducated men, but in
the process assert the superiority of theory. This new definition of the
engineer as a theoretician, someone scientifically trained who also
understood practice, became a battle cry of Newtonians like Desaguliers.
By the middle of the eighteenth century in England their vision had
won out, and with it came a new professional status for the engineer
and a new partnership with entrepreneurs. [67] A class divide had to

be bridged that required the engineers to go to workshops and industrial sites, to learn practical techniques, to unearth trade secrets, to memorize movements, to make exact drawings; in short, to better the craftsmen by applying theories to their skills. With these practical skills first British, but then increasingly Continental, engineers became key players in the trial-and-error process through which occurred the mechanization of manufacturing, transportation, and water and steam power. The Newtonians led the way in this reconceptualization of the civil engineer, but by 1800 he was a figure to be seen in almost every European country. [66]

The rise to preeminence of the British civil engineer can be seen as early as the 1720s. At the time his prowess was illustrated comparatively. The French had constructed a huge hydraulic plant at Marly; indeed, it was world famous. Desaguliers spent eight days examining it, described all fourteen of its machines, and then illustrated how some were inferior to a water mill that had been built by Henry Beighton at Barr Pool in Warwickshire or to a similar hydraulic plant on the Thames that supplied London with water. [See fig. 2.4f] While Marly used 14 wheels, the plant on the Thames used 4 and lifted three times as much water. The Marly plant cost around 4 million pounds; Desaguliers claimed that by using a Newcomen engine a similar plant could now be contructed for about 10,000 pounds. The message could not have been clearer: applied mechanics done properly translates into profitability. At the heart of the analysis, even this early, lay the steam engine. The achievement of craftsmen like Savery and Newcomen, the engine became a centerpiece in the improvements men of theory like Desaguliers promised to their customers. Likewise this new breed of engineers latched onto mechanization at factory sites. They may not have built those factories, but they would explain them and imitate the complex movements of their parts.

The cultural style exemplified by the Newtonian engineers captured the imagination of the mainstream of the British scientific establishment. The Royal Society of London, since its founding in the 1660s, had always had a utilitarian, Baconian frame upon which the "purer," more abstract science exemplified by Newton rose and flourished. But during the eighteenth century the utilitarian undergirding seemed preeminent, and so too did the generally Newtonian cast to experimental work deemed worthy. Just take the practical achievements for which the Society gave its prestigeous Copley Medal every year from the 1730s to the 1780s: experiments in electricity on several occasions; an engine for driving piles; observations on finding a northwest passage to the East Indies; resistance of the air; motion of the apparently fixed stars, measurement of time; refrangibility of light rays; water wheels and windmill sails; the air pump and condensing engine; specific gravities of metals; observations on the attraction of mountains; force of gunpowder and the velocity of cannonballs; principles of progressive and rotatory motion; and the properties of heat. There were chemical and medical topics amid the applied mechanics; but the engines and pumps largely captured the attention of the dozens of physicians, surgeons, professors, teachers, clergymen, aristocrats, gentry and gentlemen, and army and navy officers who vastly outnumbered the mere ten artisans and tradesmen who did each manage to win the award. [4]

The scientific culture being glorified by the award system of the Royal Society was overwhelmingly dominated by men of the professions, business, landed wealth, and less importantly by officers in the army and navy. By contrast and overwhelmingly, engineers in France were military men and government employees. As a result of their aristocratic status—however financially precarious the family—they had the right to train in one of the state's engineering schools and then to find a place in one or another government agency. They possessed the hauteur of their estate and the authoritarian manner of their bellicose profession. Their relations with civilians were tense; they expected obedience or at least deference. When French engineers were sent to England in the 1780s to survey roads, harbors, and bridges—perhaps even to spy—they reported back on the differences between the two societies and hence the two scientific cultures. The French military engineers marveled at the civilians who worked for and with the British navy: "The employees consider themselves to be civilians . . . and do not feel that they are inferior to the military. . . . Perhaps in France, our customs, our prejudices . . . would make this impossible to hope for although this way of thinking is certainly one of the reasons for the prosperity of the British navy." [8:145]

The French military engineers knew how difficult, if not impossible, their own relations were with civilians: "In our way of life, it is impossible for the military and civilians to work together, without the military taking over." [8:145–46] By the 1780s the French military engineers who were making these social observations had also been well trained in the applied mechanics pioneered by Desaguliers and Nollet. Their knowledge was comparable to that of British civil engineers such as John Smeaton. Indeed, they even knew and admired his work on lighthouses and waterwheels. What was different for the French military men was the social setting wherein their scientific knowledge was being utilized. The close partnerships between engineers and entrepreneurs, exemplified by the working lives of Desaguliers, Smeaton, and then Watt, were rare if not nonexistent across the Channel. [41] In France engineers excelled at massive state projects or often turned to architecture, where they transformed European styles and tastes in the direction of buildings that seemed to capture the pure light revealed in Newton's *Opticks* and that imitated the regularity and symmetry of the Newtonian universe. [66]

The closeness of British engineers and entrepreneurs arose out of the important and distinctive social setting of early Newtonianism. Forced into pulpit and lecture hall by the earlier revolutionary context wherein it was first articulated, after the *Principia* Newtonian science belonged as much to the public as it did to highly skilled practitioners. Almost effortlessly, the first Newtonians moved from being explicators to becoming civil engineers. In that move lay the cultural foundations of the British Industrial Revolution, and Newtonian science applied to industry and transportation was essential to it. [6]

Late in the eighteenth century when British industrial superiority deeply concerned French competitors, theorists and statesmen there began to search for explanations. Perhaps the shrewdest French observer of England's industrial preponderance was J. A. Chaptal (b. 1756), chemist, industrialist, ardent freemason, revolutionary, and a Napoleonic minister of the interior. [64] With constant reference to the British achievement, Chaptal sought to find the keys to industrial success. He identified mechanical and chemical knowledge, along with the division of labor, as the key elements in British entrepreneurial success in factory production.[27] In his history of French industry (*De l'industrie française,* 1819, p. 32), Chaptal explained that "in this epoque when the study of the sciences has become so general, and the relationship between *les savans et les artistes* so intimate . . . they have arrived at an extraordinary degree of perfection in their art. . . . Industry presumes on the part of the artist an extended knowledge of mechanics, various *notions de calcul,* a great dexterity in work, and enlightenment in the principles of the art." He concluded that the linkage between applied mechanics and mechanical skill had led to the perfection of various industries. Publishing what was probably the first history of industrialization Chaptal applied the term "revolution" to what had recently occurred in cotton manufacturing. Consistently he singled out the relationship between theoreticians and practitioners as the major cultural component in rapid industrial development. Had he known them as people rather than by their reputations, he could have been talking about the way engineers like Desaguliers or Smeaton or Watt and Jessop took their knowledge of mechanics to local industrial sites. In a testy partnership with entrepreneurs the visiting engineers mechanized factories, drained mines, built canals, and dredged harbors.

Early in the nineteenth century when Newtonian mechanics had become an orthodoxy throughout Europe, worried ministers reported to Napoleon on the uphill struggle faced by French industry in the midst of a fierce Continental war with the British. In a confidential report the French chose a metaphor to express their frustration. It was ironically, but correctly, Newtonian: "The absolute necessity to recreate and sustain French industry has been resolved by the English in an absolutely decisive manner. With a powerful lever they lift up an enormous mass of productions; the overwhelming gravitation of the weight [of this mass] throws everything else into its orbit."[28] Masses, weights, gravitational pull, orbits, attractions and repulsions, when incorporated into mechanics, rationalized and systematized levers, pulleys, beams, and frictions. By the 1740s the theories, laws, and devices lay at the essence of Newtonian mechanics. The equations, experiments, and gadgets were unified under the rubric of Newtonianism, and the culture that it spawned among entrepreneurs as well as engineers made unprecedented change possible. Mechanical ingenuity for the first time in human history broke through a once unbreachable barrier. Mechanical applications transformed raw human strength and the limits of unassisted human productive ability. Out of mechanization assisted by Newtonian mechanics emerged the material and mental universe—industrial and scientific—in which most Westerners and some non-Westerners now live, one aptly described as modernity.

Notes

1. Spencer Research Library, University of Kansas, Lawrence, Kansas, letter dated May 9, 1707, MS G 23, Anthony Collins probably to John Trenchard.
2. For some of Newton's political writings, see his manuscripts at King's College, Cambridge, Keynes MS 118.
3. Richard Bentley, *A Confutation of Atheism from the Origin and Frame of the World* (London: H. Mortlock, 1693), Oct. 3, 30; and in the same series the sermon of November 7, 1692, p. 28; December 5, p. 37.
4. Benjamin Martin, *A Panegyrick on the Newtonian Philosophy, shewing the Nature and Dignity of the Science . . .* (London, 1754), 5.
5. Richard Bentley, *Of Revelation and the Messias. A Sermon Preached at the Publick Commencement at Cambridge. July 5, 1696* (London, 1696), 4; and 7 and 34 for the next two quotations.
6. *Dr Bentley's Dedication of Horace, Translated . . . inscribed to the Right Honourable the Lord Halifax . . .* (London, [1712]), 19.
7. See the interesting diary begun in 1696 by Lodewijck van der Saan, secretary to the Dutch ambassador, which even contains drawings of some of these objects; University of Leiden Library, Leiden, The Netherlands, BPL 1325.
8. *A Course of Experimental Philosophy* (London, 1745), 2:412–15.

9. For example, see Cambridge University Library L 18/28, "Report of the Progress of the Students at Homerton, June 27, 1810"; and notes for 1826.

10. *Mathematical Elements of Natural Philosophy* (London, 1747), 6 ed. put out by Desaguliers's son, "An Oration concerning Evidence," xlii.

11. See Jack Fruchtman, Jr., *Thomas Paine and the Religion of Nature* (Baltimore: Johns Hopkins University Press, 1993). See also Martin Fitzpatrick, "Latitudinarianism at the Parting of the Ways: A Suggestion," in John Walsh, Colin Haydon, Stephen Taylor, eds., *The Church of England, 1688–1833* (Cambridge: Cambridge University Press, 1993), 209–27.

12. Jean Barbeyrac, "An Historical and Critical Account of the Science of Morality," preface to Samuel Pufendorf, *Of the Law of Nature and Nations,* 4th ed. (London, 1729), 4–5.

13. *An Account of the Mechanism of an Automaton, or Image playing on the German-Flute: As it was presented in a Memoire, to the Gentlemen of the Royal Academy of Sciences at Paris. By M. Vaucanson . . . ,* trans. J. T. Desaguliers (London, 1742), 9.

14. Rijkarchief Gelderland, FA van Eck MS 1041. References kindly supplied by Arianne Baggerman.

15. See, for example, Samuel Miller, *A Brief Retrospect of the Eighteenth Century* (New York, 1803).

16. See Benjamin Martin, *The General Magazine of Arts and Sciences,* part 1, vol. 1 (London, 1755); note the opening poem, "The Young Gentleman and Lady's Philsophy."

17. Algarotti, *Sir Isaac Newton's Philosophy Explained for the Use of the Ladies . . .* (London, 1739), 21–26.

18. Isaac Newton, *Opticks,* 2d ed. with additions (London, 1718), 372–73.

19. For such a person, a teacher of mathematics and officer of the excise, see the 700-page diary "Memoirs of John Cannon," Somerset Record Office, Taunton. See also the papers of John Watt from the 1720s, a mathematical practitioner and relative of James Watt; Birmingham (UK) City Library.

20. Watt manuscripts in the possession of Lord Gibson-Watt, Wales; transcribed and kindly given by Eric Robinson for our use; letter dated 5 February 1778 (Ansthorpe) to Boulton and Watt. Underlining in the original.

21. Birmingham (UK) City Library, Watt MSS, MIV/14/1, a notebook entitled "Mechanic Principles" in the hand of John Watt. For evidence that by the 1780s this educational gap in France may have been closing, see the student notebooks of Eleuthere Irenée du Pont, Hagley Museum, Longwood MSS, Series B, Box 10, course notes taken at the Collège Royal on natural history, physics, pneumatics, botany, notes on Desaguliers, Nollet, Franklin.

22. Birmingham (UK) City Library, Boulton and Watt MSS, Box 36/17, J. D. H. van Liender to Watt, 21 October 1790.

23. See Rijksarchiv Friesland, MS FA Van Sminia 1944a; the diary of Hessel Vegelin van Claerbergen.

24. William Petty, *Several Essays in Political Arithmetick* (London, 1755; orig. pub. 1682), 115.

25. See Royal Society, London, MSS CP 18, item 8, ff. 66–80, where the argument is openly made.

26. William Petty, *An Essay concerning the multiplication of Mankind* (London, 1682), 6.

27. Chaptal, *Essai sur le perfectionnement des arts chimiques en France* (Paris, 1800), 50.

28. Quoted in Archives Nationales, Paris, F12,502.

Bibliography

PART 1

Excellent reference works for students of intellectual history and the history of science are the multivolume sets *The Dictionary of the History of Ideas* and *The Dictionary of Scientific Biography.* Both will provide not only information but also additional bibliography to guide further research. For Newton himself *The Newton Handbook* by Derek Gjertsen (item [26] below) is perhaps the best starting place. All of the works listed below will also guide the student toward additional sources for research.

[1] H. G. Alexander, ed. *The Leibniz-Clarke Correspondence, Together with Extracts from Newton's "Principia" and "Opticks."* Introduction and notes by H. G. Alexander. Philosophical Classics, general ed., Peter G. Lucas. Manchester: Manchester University Press, 1956.

[2] Carl B. Boyer. *A History of Mathematics.* New York, London, and Sydney: John Wiley and Sons, 1968.

[3] Carl B. Boyer. *The History of the Calculus and Its Conceptual Development (The Concepts of the Calculus).* Foreword by Richard Courant. 1949; rpt. New York: Dover, 1959.

[4] Edwin Arthur Burtt. *The Metaphysical Foundations of Modern Science.* 2d rev. ed. Rpt. Garden City, NY: Doubleday, 1954.

[5] Max Caspar. *Kepler.* Trans. and ed. C. Doris Hellman. London: Abelard-Schuman, 1959.

[6] Ernst Cassirer. *The Platonic Renaissance in England.* Trans. James P. Pettegrove. Austin: University of Texas Press, 1953.

[7] *Catalogue of the Newton Papers Sold by Order of the Viscount Lymington to Whom They Have Descended from Catherine Conduitt, Viscountess Lymington, Great-Niece of Sir Isaac Newton.* London: Sotheby, 1936.

[8] *A Catalogue of the Portsmouth Collection of Books and Papers written by or belonging to Sir Isaac Newton, the scientific portion of which has been presented by the Earl of Portsmouth to the University of Cambridge. Drawn up by the Syndicate appointed the 6th November, 1872.* Cambridge: Cambridge University Press, 1888.

[9] Gale E. Christianson. *In the Presence of the Creator: Isaac Newton and His Times.* New York: Free Press; London: Collier Macmillan, 1984.

[10] I. Bernard. Cohen. *The Birth of a New Physics.* Rev. and updated. New York and London: Norton, 1985.

[11] I. Bernard Cohen. *Introduction to Newton's "Principia."* Cambridge: Harvard University Press; Cambridge: Cambridge University Press, 1971.

[12] I. Bernard Cohen. *The Newtonian Revolution: With Illustrations of the*

Transformation of Scientific Ideas. Cambridge and New York: Cambridge University Press, 1980.

[13] John Conduitt. "Memoirs of Sir Isaac Newton, sent by Mr. Conduitt to Monsieur Fontenelle, in 1727." In Edmund Turnor, *Collections for the History of the Town and Soke of Grantham. Containing Authentic Memoirs of Sir Isaac Newton, Now First Published From the Original MSS. in the Possession of the Earl of Portsmouth.* London: Printed for William Miller, Albemarle-Street, by W. Bulmer and Co., Cleveland-Row, St. James's, 1806.

[14] René Descartes. *Oeuvres de Descartes publiées par Charles Adam et Paul Tannery.* 11 vols. Paris: Librairie Philosophique J. Vrin, 1964–74.

[15] B. J. T. Dobbs. *Alchemical Death and Resurrection: The Significance of Alchemy in the Age of Newton. A lecture sponsored by the Smithsonian Institution Libraries in conjunction with the Washington Collegium for the Humanities Lecture Series: Death and the Afterlife in Art and Literature. Presented at the Smithsonian Institution, February 16, 1988.* Washington, DC: Smithsonian Institution Libraries, 1990.

[16] B. J. T. Dobbs. "From the Secrecy of Alchemy to the Openness of Chemistry." In Tore Frängsmyr, ed., *Solomon's House Revisited: The Organization and Institutionalization of Science. Nobel Symposium 75.* Canton, MA.: Science History Publications, 1990.

[17] B. J. T. Dobbs. *The Foundations of Newton's Alchemy, or "The Hunting of the Greene Lyon."* Cambridge and New York: Cambridge University Press, 1975.

[18] B. J. T. Dobbs. *The Janus Faces of Genius: The Role of Alchemy in Newton's Thought.* Cambridge and New York: Cambridge University Press, 1991.

[19] Samuel Y. Edgerton, Jr. *The Heritage of Giotto's Geometry: Art and Science on the Eve of the Scientific Revolution.* Ithaca and London: Cornell University Press, 1991.

[20] C. H. Edwards, Jr. *The Historical Development of the Calculus.* New York, Heidelberg, and Berlin: Springer-Verlag, 1979.

[21] John Fauvel, Raymond Flood, Michael Shortland, and Robin Wilson, eds. *Let Newton Be!* Oxford: Oxford University Press, 1988.

[22] James E. Force. *William Whiston: Honest Newtonian.* Cambridge and New York: Cambridge University Press, 1985.

[23] James E. Force and Richard H. Popkin. *Essays on the Context, Nature, and Influence of Isaac Newton's Theology.* International Archives of the History of Ideas, no. 129. Dordrecht: Kluwer Academic Publishers, 1990.

[24] LeRoy Edwin Froom. *The Prophetic Faith of Our Fathers: The Historical Development of Prophetic Interpretation.* 4 vols. Washington, DC: Review and Herald, 1946–54.

[25] Sara Schechner Genuth. "Comets, Teleology, and the Relationship of Chemistry to Cosmology in Newton's Thought." *Annali dell' Instituto e Museo di Storia della Scienza di Firenze* 10 (1985): 31–65.

[26] Derek Gjertsen. *The Newton Handbook.* London and New York: Routledge and Kegan Paul, 1986.

[27] Edward Grant. *Much Ado about Nothing: Theories of Space and Vacuum from the Middle Ages to the Scientific Revolution.* Cambridge and New York: Cambridge University Press, 1981.

[28] Robert C. Gregg and Dennis E. Groh. *Early Arianism: A View of Salvation.* Philadelphia: Fortress Press, 1981.

[29] Henry Guerlac. *Essays and Papers in the History of Modern Science.* Baltimore: Johns Hopkins University Press, 1977.

[30] Henry Guerlac. "Theological Voluntarism and Biological Analogies in Newton's Physical Thought." *Journal of the History of Ideas* 44 (1983): 219–29.

[31] Henry Guerlac and M. C. Jacob. "Bentley, Newton, and Providence (The Boyle Lectures Once More)." *Journal of the History of Ideas* 30 (1969): 307–18.

[32] David E. Hahm. *The Origins of Stoic Cosmology.* Columbus: Ohio State University Press, 1977.

[33] A. Rupert Hall. *Philosophers at War: The Quarrel between Newton and Leibniz.* Cambridge and London: Cambridge University Press, 1980.

[34] John Harrison. *The Library of Isaac Newton.* Cambridge and London: Cambridge University Press, 1978.

[35] Joan L. Hawes. "Newton and the 'Electrical Attraction Unexcited.'" *Annals of Science* 24 (1968): 121–30.

[36] Joan L. Hawes. "Newton's Revival of the Aether Hypothesis and the Explanation of Gravitational Attraction." *Notes and Records of the Royal Society of London* 23 (1968): 200–212.

[37] Joan L. Hawes. "Newton's Two Electricities." *Annals of Science* 27 (1971): 95–103.

[38] J. L. Heilbron. *Electricity in the Seventeenth and Eighteenth Centuries: A Study of Early Modern Physics.* Berkeley and Los Angeles: University of California Press, 1979.

[39] J. L. Heilbron. *Physics at the Royal Society during Newton's Presidency.* Los Angeles: William Andrews Clark Memorial Library, UCLA, 1983.

[40] John Herivel. *The Background to Newton's "Principia": A Study of Newton's Dynamical Researches in the Years 1664–84.* Oxford: Clarendon Press, 1965.

[41] W. G. Hiscock, ed. *David Gregory, Isaac Newton, and Their Circle: Extracts from David Gregory's Memoranda, 1677–1708.* Oxford: Printed for the Editor, 1937.

[42] *The Holy Bible containing the Old and New Testaments. Authorized King James Version.* Oxford: Oxford University Press, n.d.

[43] R. W. Home. "Force, Electricity, and the Powers of Living Matter in Newton's Mature Philosophy of Nature." In Margaret J. Osler and Paul Lawrence Farber, eds., *Religion, Science, and Worldview: Essays in Honor of Richard S. Westfall.* Cambridge and New York: Cambridge University Press, 1985.

[44] R. W. Home. "Newton on Electricity and the Aether." In Zev Bechler, ed., *Contemporary Newtonian Research.* Studies in the History of Modern Science, no. 9. Dordrecht: D. Reidel, 1982.

[45] Reijer Hooykaas. *Religion and the Rise of Modern Science.* Edinburgh: Scottish Academic Press, 1973.

[46] H. A. K. Hunt. *A Physical Interpretation of the Universe: The Doctrines of Zeno the Stoic.* Melbourne: Melbourne University Press, 1976.

[47] Keith Hutchinson. "Supernaturalism and the Mechanical Philosophy." *History of Science* 21 (1983): 297–333.

[48] Margaret C. Jacob. "Newton and the French Prophets: New Evidence." *History of Science* 16 (1978): 134–42.

[49] Margaret C. Jacob. *The Newtonians and the English Revolution, 1689–1720.* 1976; rpt. New York: Gordon and Breach, 1991.

[50] Margaret C. Jacob. *The Cultural Meaning of the Scientific Revolution.* 1988; new ed. New York: Oxford University Press, 1995.

[51] Robert Hugh Kargon. *Atomism in England from Hariot to Newton.* Oxford: Clarendon Press, 1966.

[52] Alexandre Koyré. *From the Closed World to the Infinite Universe.* 1957; rpt. New York: Harper and Brothers, 1958.

[53] Alexandre Koyré. *Newtonian Studies.* Cambridge: Harvard University Press, 1965.

[54] David Charles Kubrin. "Newton and the Cyclical Cosmos: Providence and the Mechanical Philosophy." *Journal of the History of Ideas* 28 (1967): 325–46.

[55] David C. Lindberg. *The Beginnings of Western Science: The European Scientific Tradition in Philosophical, Religious, and Institutional Context, 600 B.C. to A.D. 1450.* Chicago and London: University of Chicago Press, 1992.

[56] David C. Lindberg. "The Genesis of Kepler's Theory of Light: Light Metaphysics from Plotinus to Kepler." *Osiris*, 2d ser. 2 (1986): 5–42.

[57] David C. Lindberg and Ronald L. Numbers, eds. *God and Nature: Historical Essays on the Encounter between Christianity and Science.* Berkeley and Los Angeles: University of California Press, 1986.

[58] Jack Lindsay. *The Origins of Alchemy in Graeco-Roman Egypt.* New York: Barnes and Noble, 1970.

[59] A. A. Long. *Hellenistic Philosophy: Stoics, Epicureans, Sceptics.* London: Duckworth, 1974.

[60] J. E. McGuire and P. M. Heimann. "The Rejection of Newton's Concept of Matter in the Eighteenth Century." In Ernan McMullin, ed., *The Concept of Matter in Modern Philosophy.* Notre Dame and London: University of Notre Dame Press, 1978.

[61] J. E. McGuire and P. M. Rattansi. "Newton and the 'Pipes of Pan.'" *Notes and Records of the Royal Society of London* 21 (1966): 108–43.

[62] J. E. McGuire and Martin Tamny. *Certain Philosophical Questions: Newton's Trinity Notebook.* Cambridge and London: Cambridge: University Press, 1983.

[63] Ernan McMullin. *Newton on Matter and Activity.* Notre Dame: University of Notre Dame Press, 1978.

[64] Elizabeth Mackenzie. "The Growth of Plants: A Seventeenth-Century Metaphor." In *English Renaissance Studies Presented to Dame Helen Gardner in Honour of Her Seventieth Birthday.* Oxford: Clarendon Press, 1980.

[65] Frank E. Manuel. *Isaac Newton, Historian.* Cambridge: Belknap Press of Harvard University Press, 1963.

[66] Frank E. Manuel. *A Portrait of Isaac Newton.* Cambridge: Belknap Press of Harvard University Press, 1968.

[67] Frank E. Manuel. *The Religion of Isaac Newton: The Fremantle Lectures 1973.* Oxford: Clarendon Press, 1974.

[68] E. C. Millington. "Theories of Cohesion in the Seventeenth Century." *Annals of Science* 5 (1941–47): 253–69.

[69] Charles G. Nauert, Jr. *Agrippa and the Crisis of Renaissance Thought.* Illinois Studies in the Social Sciences, no. 55. Urbana: University of Illinois Press, 1965.

[70] Isaac Newton. *The Chronology of Ancient Kingdoms Amended. To which is Prefix'd, A Short Chronicle from the First Memory of Things in Europe, to the*

Conquest of Persia by Alexander the Great. London: Printed for J. Tonson in the Strand, and J. Osborn and T. Longman in Paternoster Row, 1728.

[71] Isaac Newton. *The Correspondence of Isaac Newton.* Ed. H. W. Turnbull, J. P. Scott, A. R. Hall, and Laura Tilling. 7 vols. Cambridge: Published for the Royal Society at the University Press, 1959–77.

[72] Isaac Newton. *Manuscripts and Papers Collected and Published on Microfilm by Chawyck-Healey.* Ed. Peter Jones. Cambridge: Chadwyck-Healey, 1991.

[73] Isaac Newton. *The Mathematical Papers of Isaac Newton.* Ed. Derek T. Whiteside with the assistance in publication of M. A. Hoskin. 8 vols. Cambridge: Cambridge University Press, 1967–80.

[74] Isaac Newton. *Observations upon the Prophecies of Daniel, and the Apocalypse of St. John. In Two Parts.* London: Printed by J. Darby and T. Browne in Bartholomew-Close. And sold by J. Roberts in Warwick-lane, J. Tonson in the Strand, W. Innys and R. Manby at the West End of St. Paul's Church-Yard, J. Osborn and T. Longman in Pater-Noster-Row, J. Noon near Mercers Chapel in Cheapside, T. Hatchett at the Royal Exchange, S. Harding in St. Martin's lane, J. Stagg in Westminster-Hall, J. Parker in Pall-mall, and J. Brindley in New Bond-street, 1733.

[75] Isaac Newton. *Opticks, or A Treatise of the Reflections, Refractions, Inflections & Colours of Light.* Foreword by Albert Einstein, introduction by Sir Edmund Whittaker, preface by I. Bernard Cohen, analytical table of contents by Duane H. D. Roller. Based on the 4th London ed. of 1730. New York: Dover, 1952.

[76] Isaac Newton. *Sir Isaac Newton: Theological Manuscripts.* Selected and ed. with an introduction by H. McLachlan. Liverpool: At the University Press, 1950.

[77] Isaac Newton. *Sir Isaac Newton's Mathematical Principles of Natural Philosophy and His System of the World.* Trans. Andrew Motte, 1729. Ed. Florian Cajori. 2 vols. 1934; rpt. Berkeley and Los Angeles: University of California Press, 1962.

[78] Isaac Newton. *Unpublished Scientific Papers of Isaac Newton: A Selection from the Portsmouth Collection in the University Library, Cambridge. Chosen, edited, and translated by A. Rupert Hall and Marie Boas Hall.* Cambridge: Cambridge University Press, 1962.

[79] Margaret J. Osler. "Descartes and Charleton on Nature and God." *Journal of the History of Ideas* 40 (1979): 445–56.

[80] Margaret J. Osler. "Eternal Truths and the Laws of Nature: The Theological Foundations of Descartes' Philosophy of Nature." *Journal of the History of Ideas* 46 (1985): 349–62.

[81] Margaret J. Osler. "Providence and Divine Will in Gassendi's Views on Scientific Knowledge." *Journal of the History of Ideas* 44 (1983): 549–60.

[82] Edward Peters, ed. *Heresy and Authority in Medieval Europe: Documents in Translation.* Introduction by Edward Peters. Philadelphia: University of Pennsylvania Press, 1980.

[83] Richard H. Popkin. *The History of Scepticism from Erasmus to Spinoza.* Berkeley and Los Angeles: University of California Press, 1979.

[84] Richard H. Popkin, ed. *Millenarianism and Messianism in English Literature and Thought, 1650–1800. Clark Library Lectures, 1981–82.* Publications from the Clark Library Professorship, UCLA, no. 10. Leiden: E. J. Brill, 1988.

[85] Arthur Quinn. *The Confidence of British Philosophers: An Essay in Historical*

Narrative. Studies in the History of Christian Thought, vol. 17. Ed. Heiko A. Oberman, in cooperation with Henry Chadwick, Edward A. Dowey, Jaroslav Pelikan, and E. David Willis. Leiden: E. J. Brill, 1977.

[86] John Rist, ed. *The Stoics.* Berkeley and Los Angeles: University of California Press, 1978.

[87] Danton B. Sailor. "Newton's Debt to Cudworth." *Journal of the History of Ideas* 49 (1988): 511–18.

[88] Samuel Sambursky. *Physics of the Stoics.* 1959; rpt. London: Hutchinson, 1971.

[89] Samuel Sandmel. *Philo of Alexandria: An Introduction.* New York: Oxford University Press, 1979.

[90] Jason Lewis Saunders. *Justus Lipsius: The Philosophy of Renaissance Stoicism.* New York: Liberal Arts Press, 1955.

[91] Simon Schaffer. "Newton's Comets and the Transformation of Astrology." In Patrick Curry, ed., *Astrology, Science and Society. Historical Essays.* Woodbridge, Suffolk: Boydell Press, 1987.

[92] P. B. Scheuer and G. Debrock, eds. *Newton's Scientific and Philosophical Legacy.* International Archives of the History of Ideas, no. 123. Dordrecht: Kluwer Academic Publishers, 1988.

[93] Hillel Schwartz. *The French Prophets: The History of a Millenarian Group in Eighteenth-Century England.* Berkeley and Los Angeles: University of California Press, 1980.

[94] Taylor, Frank Sherwood. "The Idea of the Quintessence." In E. Ashworth Underwood, ed., *Science, Medicine, and History: Essays on the Evolution of Scientific Thought and Medical Practice Written in Honour of Charles Singer,* vol. 1. London: Oxford University Press, Geoffrey Cumberlege, 1953.

[95] Arnold Thackray. *Atoms and Powers: An Essay on Newtonian Matter-Theory and the Development of Chemistry.* Cambridge: Harvard University Press, 1970.

[96] G. Verbeke. *L'Evolution de la doctrine du pneuma du stoicism à S. Augustin: Etude Philosophique.* Bibliothèque de l'Institut Supérieur de Philosophie Université de Louvain. Paris: Desclée De Brouwer; Louvain: Editions de l'Institut Supérieur de Philosophie, 1945.

[97] Jacob Viner. *The Role of Providence in the Social Order: An Essay in Intellectual History. Jayne Lectures for 1966.* Foreword by Joseph R. Strayer. Memoirs of the American Philosophical Society Held at Philadelphia for Promoting Useful Knowledge, vol. 90. Philadelphia: American Philosophical Society, 1972.

[98] W. A. Wallace. "Newton's Early Writings: Beginning of a New Direction." In G. V. Coyne, M. Heller, and J. Zyciński, eds., *Newton and the New Direction in Science: Proceedings of the Cracow Conference 25 to 28 May 1987.* Vatican City: Specola Vaticana, 1988.

[99] Charles Webster *From Paracelsus to Newton: Magic and the Making of Modern Science. The Eddington Memorial Lectures Delivered at Cambridge, November 1980.* Cambridge: Cambridge University Press, 1982.

[100] Richard S. Westfall. "Isaac Newton's *Theologiae Gentilis Origines Philosophicae.*" In W. Warren Wagar, ed., *The Secular Mind: Transformations of Faith in Modern Europe. Essays Presented to Franklin L. Baumer, Randolph W. Townsend Professor of History, Yale University.* New York: Holmes and Meier, 1982.

[101] Richard S. Westfall. *Never at Rest: A Biography of Isaac Newton.* Cambridge and New York: Cambridge University Press, 1980.

[102] Derek T. Whiteside. "Before the *Principia*: The Maturing of Newton's Thoughts on Dynamical Astronomy, 1664–1684." *Journal for the History of Astronomy* 1 (1970): 5–19.

[103] Derek T. Whiteside. "Isaac Newton: Birth of a Mathematician." *Notes and Records of the Royal Society of London* 19 (1964): 53–62.

[104] Derek T. Whiteside. "Sources and Strengths of Newton's Early Mathematical Thought." In *The Annus Mirabilis of Sir Isaac Newton Tricentennial Celebration. Texas Quarterly* 10, no. 3 (Autumn 1967): 69–85.

[105] Lancelot Law Whyte. *Essay on Atomism: From Democritus to 1960.* 1961; rpt. New York: Harper and Row, 1963.

[106] Curtis Wilson. "How Did Kepler Discover His First Two Laws?" *Scientific American* 226, no. 3 (1972): 93–106.

[107] Curtis Wilson. "Kepler's Derivation of the Elliptical Path." *Isis* 59 (1968): 5–25.

PART 2

For students not very familiar with the intellectual and cultural history of the eighteenth century, there are a few good general books with which to begin: Roy Porter, *The Enlightenment* (Atlantic Highlands, N.J.: Humanities Press, 1991); Thomas L. Hankins, *Science and the Enlightenment* (Cambridge: Cambridge University Press, 1985); John Hedley Brooke, *Science and Religion: Some Historical Perspectives* (Cambridge: Cambridge University Press, 1991). For individual Newtonians, the *Dictionary of National Biography* (DNB) can also be consulted.

[1] Andrew B. Appleby. "Grain Prices and Subsistence Crises in England and France, 1590–1740." *Journal of Economic History* 39 (1979): 865–87.

[2] James L. Axtell. "Locke, Newton, and the Two Cultures," in J. Yolton, ed., *John Locke: Problems and Perspectives.* Cambridge: Cambridge University Press, 1969.

[3] S. A. Bedini. "Of 'Science and Liberty': The Scientific Instruments of King's College and Eighteenth-Century Columbia College in New York." *Annals of Science* 50 (1993): 201–28.

[4] M. Yakup Bektas and Maurice Crosland. "The Copley Medal: The Establishment of a Reward System in the Royal Society, 1731–1839." *Notes and Records of the Royal Society of London* 46 (1992): 43–76.

[5] Maxine Berg. "Women's Work, Mechanization, and the Early Phases of Industrialisation in England," in P. Joyce, ed. *The Historical Meanings of Work.* Cambridge: University of Cambridge Press, 1987.

[6] Maxine Berg and Pat Hudson. "Rehabilitating the Industrial Revolution." *Economic History Review* 45 (1992): 24–50.

[7] Richard Biernacki. *The Fabrication of Labor in Germany and Britain, 1640–1914.* Berkeley and Los Angeles: University of California Press, forthcoming.

[8] Margaret Bradley. "Engineers as Military Spies? French Engineers Come to Britain, 1780–1790." *Annals of Science* 49 (1992) 137–61.

[9] L. W. B. Brockliss. *French Higher Education in the Seventeenth and Eighteenth Centuries.* Oxford: Clarendon Press, 1987.

[10] Geoffrey Bowles. "The Place of Newtonian Explanation in English Popular Thought." Ph.D. diss., Oxford University, 1977.

[11] F. Forrester Church. "The Gospel According to Thomas Jefferson," in his edition of *The Jefferson Bible*. Boston: Beacon Press, 1989.

[12] J. C. D. Clark. *Revolution and Rebellion: State and Society in England in the Seventeenth and Eighteenth Centuries*. Cambridge: Cambridge University Press, 1986.

[13] James S. Coleman. *Foundations of Social Theory*. Cambridge: Harvard University Press, 1990.

[14] Andrew Cunningham. "How the *Principia* Got Its Name." *History of Science* 29 (1991): 377–92.

[15] C. A. Davids. *Zeewezen en wetenschap: De wetenschap en de ontwikkeling van de navigatie techniek in Nederland tussen 1585 en 1815*. Amsterdam, 1986.

[16] Eamon Duffy. " 'Whiston's Affair': The Trials of a Primitive Christian, 1709–1714." *Journal of Ecclesiastical History* 27 (1976): 129–50.

[17] Raymond J. Evans. "The Diffusion of Science: The Geographical Transmission of Natural Philosophy into the English Provinces, 1660–1760." Ph.D. diss., Cambridge University Library, 1982.

[18] Paula Findlen. "Science as a Career in Enlightenment Italy: The Strategies of Laura Bassi." *Isis* 84 (1993): 441–69.

[19] James E. Force and Richard H. Popkin. *Essays on the Context, Nature, and Influence of Isaac Newton's Theology*. International Archives of the History of Ideas, no. 129. Dordrecht: Kluwer Academic Publishers, 1990.

[20] James E. Force. *William Whiston: Honest Newtonian*. Cambridge and New York: Cambridge University Press, 1985.

[21] Gad Freudenthal. "Early Electricity between Chemistry and Physics: The Simultaneous Itineraries of Francis Hauksbee, Samuel Wall, and Pierre Poliniere." *Historical Studies in the Physical Sciences* 11 (1982): 203–29.

[22] John Gascoigne. *Cambridge in the Age of the Enlightenment: Science, Religion, and Politics from the Restoration to the French Revolution*. Cambridge: Cambridge University Press, 1989.

[23] Robert Gascoigne. "The Historical Demography of the Scientific Community, 1450–1900." *Social Studies of Science* 22 (1992): 545–73.

[24] Brian Gee. "Amusement Chests and Portable Laboratories: Practical Alternatives to the Regular Laboratory," in Frank A. J. L. James, ed., *The Development of the Laboratory: Essays on the Place of Experiment in Industrial Civilization*. London: Macmillan, 1989.

[25] Neal C. Gillespie. "Natural History, Natural Theology, and Social Order: John Ray and the 'Newtonian Ideology.' " *Journal of the History of Biology* 20 (1987): 1–49.

[26] Neal C. Gillespie. "William Paley and Divine Design." *Isis* 81 (1990): 214–29.

[27] Jan V. Golinski. "A Noble Spectacle: Phosphorous and the Public Culture of Science in the Early Royal Society." *Isis* 80 (1989): 11–39.

[28] Richard G. Griffiths. *Industrial Retardation in the Netherlands, 1830–50*. The Hague: Nijhoff, 1979.

[29] Henry Guerlac. *Newton on the Continent*. Ithaca: Cornell University Press, 1981.

[30] Anita Guerrini. "James Keill, George Cheyne, and Newtonian Physiology, 1690–1740." *Journal of the History of Biology* 18 (1985): 247–66.

[31] A. Rupert Hall. "Engineering and the Scientific Revolution." *Technology and Culture* 2, no. 4, (1961): 330–40.

[32] A. Rupert Hall. "Further Newton Correspondence." *Notes and Records of the Royal Society of London* 37, no. 1 (1982): 7–34.

[33] Nicholas Hans. *New Trends in Education in the Eighteenth Century*. London: Routledge and Kegan Paul, 1966.

[34] J. L. Heilbron. *Physics at the Royal Society During Newton's Presidency*. Los Angeles: William Andrews Clark Memorial Library, UCLA, 1983.

[35] Christopher Hill. *The English Bible and the Seventeenth-Century Revolution*. London: Allen Lane, 1993.

[36] Richard L. Hills. *Power from Steam: A History of the Stationary Steam Engine*. Cambridge: Cambridge University Press, 1989.

[37] Mary Holbrook with R. G. W. Anderson and D. J. Bryden. *Science Preserved: A Directory of Scientific Instruments in Collections in the United Kingdom and Eire*. London: Science Museum, 1992.

[38] Jonathan Irvine Israel. *Dutch Primacy in World Trade, 1585–1740*. New York: Oxford University Press, 1989.

[39] Ian Inkster. "The Public Lecture as an Instrument of Science Education for Adults: The Case of Great Britain, c. 1750–1850." *Paedogogica historica* 20 (1981): 80–112.

[40] James R. Jacob. "The Political Economy of Science in Seventeenth-Century England." *Social Research* 59 (1992): 505–32. Rpt. in [44].

[41] Margaret C. Jacob. *The Cultural Meaning of the Scientific Revolution*. 1988; new ed., New York: Oxford University Press, 1995.

[42] Margaret C. Jacob. *Living the Enlightenment: Freemasonry and Politics in Eighteenth-Century Europe*. New York: Oxford University Press, 1991.

[43] Margaret C. Jacob. *The Newtonians and the English Revolution*. 1976; rpt. New York: Gordon and Breach, 1991.

[44] Margaret C. Jacob, ed. *The Politics of Western Science 1640–1990*. Atlantic Highlands, NJ: Humanities Press, 1994.

[45] Margaret C. Jacob. "Radicalism in the Dutch Enlightenment," in Margaret C. Jacob and Wijnand Mijnhardt, eds., *The Dutch Republic in the Eighteenth Century: Decline, Enlightenment, and Revolution*. Ithaca: Cornell University Press, 1992.

[46] Stephen Nicholas Jolley. "Leibniz's Critique of Locke with Special Reference to Metaphysical and Theological Themes." Ph.D. diss., Cambridge University Library, 1974.

[47] K. S. Kirsanov. "The Earliest Copy in Russia of Newton's *Principia*: Is It David Gregory's Annotated Copy?" *Notes and Records of the Royal Society of London* 46, no. 2 (1992): 203–18.

[48] Joseph M. Levine. *Dr. Woodward's Shield: History, Science, and Satire in Augustan England*. Berkeley and Los Angeles: University of California Press, 1977.

[49] L. M. Lomüller. *Guillaume Ternaux, 1763–1833: Createur de la première intégration industrielle française*. Académie nationale de Reims: Les Editions de la Cabro d'Or, Paris, 1977.

[50] Christine MacLeod. *Inventing the Industrial Revolution: The English Patent System, 1660–1800*. Cambridge: Cambridge University Press, 1988.

[51] James E. McClellan III. *Science Reorganized: Scientific Societies in the Eighteenth Century*. New York: Columbia University Press, 1985.

[52] Neil McKendrick, John Brewer, and J. H. Plumb. *The Birth of a Consumer Society: The Commercialization of Eighteenth-Century England*. Bloomington: Indiana University Press, 1982.

[53] Domenico Bertoloni Meli. "St. Peter and the Rotation of the Earth: The Problem of the Fall around 1800," in P. M. Harman and Alan E. Shapiro, eds., *The Investigation of Difficult Things: Essays on Newton and the History of the Exact Sciences.* Cambridge: Cambridge University Press, 1992.

[54] John R. Millburn. *Benjamin Martin: Author, Instrument-Maker, and "Country Showman."* Leiden: Noordhoff, 1976.

[55] John R. Millburn. "The London Evening Courses of Benjamin Martin and James Ferguson, Eighteenth-Century Lecturers on Experimental Philosophy." *Annals of Science* 40 (1983): 437–55.

[56] David P. Miller. " 'Into the Valley of Darkness': Reflections on the Royal Society in the Eighteenth Century." *History of Science* 27 (1989): 155–66.

[57] Jack Morrell and Arnold Thackray. *Gentlemen of Science: Early Years of the British Association for the Advancement of Science.* Oxford: Clarendon Press, 1981.

[58] A. Q. Morton. "Lectures on Natural Philosophy in London, 1750–1765: S. C. T. Demainbray (1710–1782) and the 'Inattention' of His Countrymen." *British Journal for the History of Science* 23 (1990): 411–34.

[59] A. E. Musson and Eric Robinson. *Science and Technology in the Industrial Revolution.* Rpt. New York: Gordon and Breach, 1991.

[60] Ronald L. Numbers. *Creation by Natural Law: Laplace's Nebular Hypothesis in American Thought.* Seattle: University of Washington Press, 1977.

[61] Kathleen H. Ochs. "The Failed Revolution in Applied Science: Studies of Industry by Members of the Royal Society of London, 1660–1688." Ph.D. diss., University of Toronto, 1981.

[62] Kathleen H. Ochs. "The Royal Society of London's History of Trades Programme: An Early Episode in Applied Science." *Notes and Records of the Royal Soceity* 39 (1985): 129–58.

[63] Richard G. Olson. "Tory–High Church Opposition to Science and Scientism in the Eighteenth Century: The Works of John Arbuthnot, Jonathan Swift, and Samuel Johnson," in John G. Burke, ed., *The Uses of Science in the Age of Newton.* Berkeley and Los Angeles: University of California Press, 1983.

[64] Michel Pérronnet, ed. *Chaptal.* Paris: Bibliothèque historique Privat, 1988.

[65] O. S. Pickering. "The Re-Emergence of the Engineering Reports of John Grundy of Spalding (1719–83)." *Transactions of the Newcomen Society,* 1988–89 (pub. 1991), 137–43.

[66] Antoine Picon. *French Architects and Engineers in the Age of Enlightenment,* trans. Martin Thom. Cambridge: Cambridge University Press, 1988.

[67] Carlo Poni. "The Craftsman and the Good Engineer: The Technical Practice and Theoretical Mechanics in J. T. Desaguliers." *History and Technology* 10 (1993): 215–32.

[68] Roy Porter. "Science, Provincial Culture, and Public Opinion in Enlightenment England." *British Journal for Eighteenth-Century Studies* 3 (1980): 20–46.

[69] L. T. C. Rolt and J. S. Allen. *The Steam Engine of Thomas Newcomen.* New York: Science History Publications, 1977.

[70] G. S. Rousseau. "Science Books and Their Readers in the Eighteenth Century," in Isabel Rivers, ed., *Books and Their Readers in Eighteenth-Century England.* Leicester: Leicester University Press, 1982.

[71] David B. Ruderman. "Jewish Thought in Newtonian England: The Career

and Writings of David Nieto." *Proceedings of the American Academy for Jewish Research.* 58 (1992): 193–219.

[72] Gordon Rupp. *Religion in England, 1688–1791.* Oxford: Clarendon Press, 1986.

[73] Simon Schaffer. "Authorized Prophets: Comets and Astronomers after 1795." *Studies in Eighteenth-Century Culture* 17 (1987): 45–74.

[74] Simon Schaffer. "Natural Philosophy and Public Spectacle in the Eighteenth Century." *History of Science* 21 (1983): pp. 1–43.

[75] Steven Shapin. "Of Gods and Kings: Natural Philosophy and Politics in the Leibniz-Clarke Disputes." *Isis* 72 (1981): 187–215.

[76] Larry Stewart. *The Rise of Public Science: Rhetoric, Technology, and Natural Philosophy in Newtonian Britain, 1660–1750.* Cambridge: Cambridge University Press, 1992.

[77] Larry Stewart. "Samuel Clarke, Newtonianism, and the Factions of Post-Revolutionary England." *Journal of the History of Ideas* 43 (1981): 53–72.

[78] Rick Szostak. *The Role of Transportation in the Industrial Revolution: Comparison of England and France.* Montreal: McGill–Queen's University Press, 1991.

[79] E. G. R. Taylor. *The Mathematical Practitioners of Hanoverian England.* Cambridge: Cambridge University Press, 1954.

[80] Deborah Valenze. "The Art of Women and the Business of Men: Women's Work and the Dairy Industry, c. 1740–1840." *Past and Present* 130 (1991): 142–69.

[81] J. Van den Berg. "Eighteenth-Century Dutch Translations of the Works of some British Latitudinarians and Enlightened Theologians." *Nederlands archief voor kerkgeschiedenis,* n.s. 59, no. 2 (1979): 198–204.

[82] Wyger R. E. Velema. *Enlightenment and Conservatism in the Dutch Republic: The Political Thought of Elie Luzac (1721–96).* Maastricht: Van Gorcum, 1993.

[83] Peter J. Wallis. "Dissemination of Mathematics in the Eighteenth Century: The Evidence of Subscription Lists." *Project for Historical Bibliography* no. 314, December 1982.

[84] P. J. Wallis. and F. J. G. Robinson. "Scientists and Subscription Lists." *British Journal for the History of Science* 6 (1972): 227–28.

[85] H. J. J. Winter, "Scientific Associations of the Spalding Gentlemen's Society during the Period 1710–50." *Archives internationale d'histoire des sciences* 3 (1950): 77–88.

[86] Kathleen Wellman. "Medicine as the Key to Defining Enlightenment Issues: The Case of Julien Offray de La Mettrie." *Studies in Eighteenth-Century Culture* 17 (1987): 75–89.

[87] Douglas L. Wilson. "Thomas Jefferson's Library and the French Connection." *Eighteenth-Century Studies* 26 (1993): 669–85.

[88] John P. Wright. "Boerhaave on Minds, Human Beings, and Mental Diseases." *Studies in Eighteenth-Century Culture* 20 (1990): 289–302.

Index